Amino Acids: Biochemistry and Applications

Amino Acids: Biochemistry and Applications

Edited by
Fraser Barrett

☐ Larsen & Keller
www.larsen-keller.com

Amino Acids: Biochemistry and Applications
Edited by Fraser Barrett
ISBN: 978-1-63549-024-4 (Hardback)

▤ Larsen & Keller

Published by Larsen and Keller Education,
5 Penn Plaza,
19th Floor,
New York, NY 10001, USA

Cataloging-in-Publication Data

Amino acids : biochemistry and applications / edited by Fraser Barrett
       p. cm.
Includes bibliographical references and index.
ISBN 978-1-63549-024-4
1. Amino acids. 2. Amino compounds. 3. Organic acids. 4. Biochemistry.
I. Barrett, Fraser.
QD431 .A45 2017
572.65--dc23

The publisher's policy is to use permanent paper from mills that operate a sustainable forestry policy. Furthermore, the publisher ensures that the text paper and cover boards used have met acceptable environmental accreditation standards.

Printed and bound in the United States of America.

For more information regarding Larsen and Keller Education and its products, please visit the publisher's website www.larsen-keller.com

Table of Contents

Preface   **VII**

Chapter 1 **Introduction to Amino Acid**   **1**
    a. Amino Acid   1
    b. Amine   18
    c. Carboxylic Acid   27
    d. Amino Acid Dating   35
    e. Nullomers   37

Chapter 2 **Essential Amino Acid: An Overview**   **41**
    a. Essential Amino Acid   41
    b. Histidine   45
    c. Isoleucine   50
    d. Leucine   52
    e. Lysine   56
    f. Methionine   62
    g. Phenylalanine   70
    h. Threonine   74
    i. Tryptophan   77
    j. Valine   83

Chapter 3 **Various Types of Amino Acid**   **86**
    a. Glutamic Acid   86
    b. Proteinogenic Amino Acid   90
    c. Non-Proteinogenic Amino Acids   104
    d. Beta-Alanine   114
    e. Amino Acid Neurotransmitter   115
    f. Alpha-Aminobutyric Acid   117
    g. Gamma-Aminobutyric Acid   117
    h. Carnitine   123
    i. Citrulline   127
    j. Domoic Acid   128

Chapter 4 **Chemical Synthesis of Amino Acids**   **133**
    a. Amino Acid Synthesis   133
    b. Peptide Synthesis   143
    c. Arndt–Eistert Reaction   158
    d. Corey–Link Reaction   159
    e. Petasis Reaction   166
    f. Schöllkopf Method   179
    g. Strecker Amino Acid Synthesis   180
    h. Enantioselective Synthesis   182

Chapter 5   **Expanded Genetic Code: An Integrated Study**   **192**
   a. Expanded Genetic Code   192
   b. Genetic Code   203
   c. Transfer RNA   214
   d. Stop Codon   221

Permissions

Index

Preface

Amino acids are organic compounds found in our bodies which contain carboxylic acid (-COOH), amines and side chains. They mainly contain hydrogen, carbon, nitrogen and oxygen. These acids are very important for nutrition and are therefore, used in food technologies and nutrition supplements. After water, amino acids comprise the largest component in human cells and tissues. This book presents the complex subject of amino acids in the most comprehensible and easy to understand language. The topics included in it are of utmost significance and are bound to provide incredible insights to readers. Different approaches, evaluations and methodologies on amino acids have been included in this text. It is a complete source of knowledge on the present status of this important field.

Given below is the chapter wise description of the book:

Chapter 1- Amino acids are organic compounds that have key elements such as carbon, hydrogen and oxygen. The number of amino acids that are known to us are 500 and they can be classified in many ways. They play an essential role in various parts of the body. The chapter is an overview of the subject matter incorporating all the major aspects of amino acids.

Chapter 2- An essential amino acid is an amino acid that cannot be created by an organism from scratch. Histidine, isoleucine, leucine, methionine, tryptophan and valine are some of the topics explained in the text. The text strategically encompasses and incorporates the major components and key concepts of amino acids, providing a complete understanding.

Chapter 3- Acids which are very important for metabolism are found abundantly in nature and have to be consumed for bodily utilization. The various types of amino acids discussed are glutamic acid, proteinogenic amino acid, non-proteinogenic amoino acids, beta-Alanie, gamma-aminobutyric acid, citulline etc. The major categories of amino acids are dealt with great details in this section.

Chapter 4- Amino acids can be synthesized by a number of ways. Amino acid synthesis is a set of processes that is produced by other compounds. Peptide synthesis, Arndt-Eistert reaction, Corey-Link reaction, Miller-Urey experiment and Strecker amino acid synthesis are some of the themes elucidated in the following section. The topics discussed in the chapter are of great importance to broaden the existing knowledge on amino acids.

Chapter 5- Expanded genetic codes are unnaturally altered genetic codes. Expanding genetic code is an interdisciplinary subject of synthetic biology. The important topics related to expanded genetic coding are peptide synthesis, genetic code, transfer RNA and stop codon. This chapter elucidates the crucial theories and principles relater to expanded genetic code.

Indeed, my job was extremely crucial and challenging as I had to ensure that every chapter is informative and structured in a student-friendly manner. I am thankful for the support provided by my family and colleagues during the completion of this book.

Editor

Introduction to Amino Acid

Amino acids are organic compounds that have key elements such as carbon, hydrogen and oxygen. The number of amino acids that are known to us are 500 and they can be classified in many ways. They play an essential role in various parts of the body. The chapter is an overview of the subject matter incorporating all the major aspects of amino acids.

Amino Acid

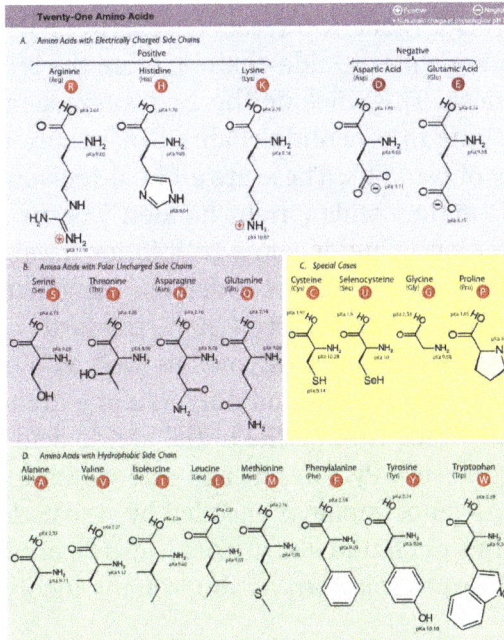

The 21 proteinogenic α-amino acids found in eukaryotes, grouped according to their side-chains' pK_a values and charges carried at physiological pH 7.4

Amino acids are biologically important organic compounds containing amine (-NH$_2$) and carboxylic acid (-COOH) functional groups, usually along with a side-chain (R group) specific to each amino acid. The key elements of an amino acid are carbon, hydrogen, oxygen, and nitrogen, though other elements are found in the side-chains of certain amino acids. About 500 amino acids are known (though only 20 appear in the genetic code) and can be classified in many ways. They can be classified according to the core structural functional groups' locations as alpha- (α-), beta- (β-), gamma- (γ-) or delta- (δ-) amino acids; other categories relate to polarity, pH level, and side-chain group type (aliphatic, acyclic, aromatic, containing hydroxyl or sulfur, etc.). In the form of proteins, amino acids comprise the second-largest component (water is the largest) of human muscles, cells and other tissues. Outside proteins, amino acids perform critical roles in processes such as neurotransmitter transport and biosynthesis.

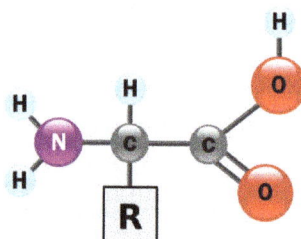

The structure of an alpha amino acid in its un-ionized form

In biochemistry, amino acids having both the amine and the carboxylic acid groups attached to the first (alpha-) carbon atom have particular importance. They are known as 2-, alpha-, or α-amino acids (generic formula $H_2NCHRCOOH$ in most cases, where R is an organic substituent known as a "side-chain"); often the term "amino acid" is used to refer specifically to these. They include the 23 proteinogenic ("protein-building") amino acids, which combine into peptide chains ("polypeptides") to form the building-blocks of a vast array of proteins. These are all L-stereoisomers ("left-handed" isomers), although a few D-amino acids ("right-handed") occur in bacterial envelopes, as a neuromodulator (D-serine), and in some antibiotics. Twenty of the proteinogenic amino acids are encoded directly by triplet codons in the genetic code and are known as "standard" amino acids. The other three ("non-standard" or "non-canonical") are selenocysteine (present in many noneukaryotes as well as most eukaryotes, but not coded directly by DNA), pyrrolysine (found only in some archea and one bacterium) and N-formylmethionine (which is often the initial amino acid of proteins in bacteria, mitochondria, and chloroplasts). Pyrrolysine and selenocysteine are encoded via variant codons; for example, selenocysteine is encoded by stop codon and SECIS element. Codon–tRNA combinations not found in nature can also be used to "expand" the genetic code and create novel proteins known as alloproteins incorporating non-proteinogenic amino acids.

Many important proteinogenic and non-proteinogenic amino acids also play critical non-protein roles within the body. For example, in the human brain, glutamate (standard glutamic acid) and gamma-amino-butyric acid ("GABA", non-standard gamma-amino acid) are, respectively, the main excitatory and inhibitory neurotransmitters; hydroxyproline (a major component of the connective tissue collagen) is synthesised from proline; the standard amino acid glycine is used to synthesise porphyrins used in red blood cells; and the non-standard carnitine is used in lipid transport.

Nine proteinogenic amino acids are called "essential" for humans because they cannot be created from other compounds by the human body and so must be taken in as food.

Others may be conditionally essential for certain ages or medical conditions. Essential amino acids may also differ between species.

Because of their biological significance, amino acids are important in nutrition and are commonly used in nutritional supplements, fertilizers, and food technology. Industrial uses include the production of drugs, biodegradable plastics, and chiral catalysts.

History

The first few amino acids were discovered in the early 19th century. In 1806, French chemists Louis-Nicolas Vauquelin and Pierre Jean Robiquet isolated a compound in asparagus that was subsequently named asparagine, the first amino acid to be discovered. Cystine was discovered in 1810, although its monomer, cysteine, remained undiscovered until 1884. Glycine and leucine were discovered in 1820. The last of the 20 common amino acids to be discovered was threonine in 1935 by William Cumming Rose, who also determined the essential amino acids and established the minimum daily requirements of all amino acids for optimal growth.

Usage of the term *amino acid* in the English language is from 1898. Proteins were found to yield amino acids after enzymatic digestion or acid hydrolysis. In 1902, Emil Fischer and Franz Hofmeister proposed that proteins are the result of the formation of bonds between the amino group of one amino acid with the carboxyl group of another, in a linear structure that Fischer termed "peptide".

General Structure

In the structure shown at the top of the page, R represents a side-chain specific to each amino acid. The carbon atom next to the carboxyl group (which is therefore numbered 2 in the carbon chain starting from that functional group) is called the α–carbon. Amino acids containing an amino group bonded directly to the alpha carbon are referred to as *alpha amino acids*. These include amino acids such as proline which contain secondary amines, which used to be often referred to as "imino acids".

Isomerism

The two enantiomers of alanine, D-alanine and L-alanine

The alpha amino acids are the most common form found in nature, but only when occurring in the L-isomer. The alpha carbon is a chiral carbon atom, with the exception of glycine which has two indistinguishable hydrogen atoms on the alpha carbon. Therefore, all alpha amino acids but glycine can exist in either of two enantiomers, called L or D amino acids, which are mirror images of each other. While L-amino acids represent all of the amino acids found in proteins during translation in the ribosome, D-amino acids are found in some proteins produced by enzyme posttranslational modifications after translation and translocation to the endoplasmic reticulum, as in exotic sea-dwelling organisms such as cone snails. They are also abundant components of the peptidoglycan cell walls of bacteria, and D-serine may act as a neurotransmitter in the brain. D-amino acids are used in racemic crystallography to create centrosymmetric crystals, which (depending on the protein) may allow for easier and more robust protein structure determination. The L and D convention for amino acid configuration refers not to the optical activity of the amino acid itself but rather to the optical activity of the isomer of glyceraldehyde from which that amino acid can, in theory, be synthesized (D-glyceraldehyde is dextrorotatory; L-glyceraldehyde is levorotatory). In alternative fashion, the *(S)* and *(R)* designators are used to indicate the absolute stereochemistry. Almost all of the amino acids in proteins are *(S)* at the α carbon, with cysteine being *(R)* and glycine non-chiral. Cysteine has its side-chain in the same geometric position as the other amino acids, but the *R/S* terminology is reversed because of the higher atomic number of sulfur compared to the carboxyl oxygen gives the side-chain a higher priority, whereas the atoms in most other side-chains give them lower priority.

Side Chains

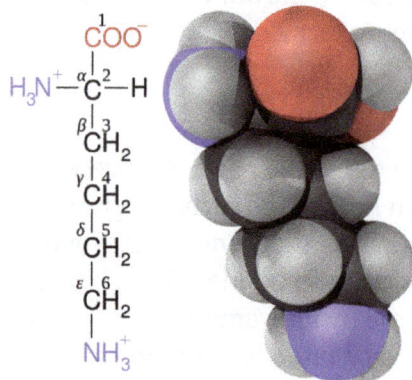

Lysine with the carbon atoms in the side-chain labeled

In amino acids that have a carbon chain attached to the α–carbon the carbons are labeled in order as α, β, γ, δ, and so on. In some amino acids, the amine group is attached to the β or γ-carbon, and these are therefore referred to as *beta* or *gamma amino acids*.

Amino acids are usually classified by the properties of their side-chain into four groups. The side-chain can make an amino acid a weak acid or a weak base, and a hydrophile if the side-chain is polar or a hydrophobe if it is nonpolar. The chemical structures of the 22 standard amino acids, along with their chemical properties, are described more fully in the article on these proteinogenic amino acids.

The phrase "branched-chain amino acids" or BCAA refers to the amino acids having aliphatic side-chains that are non-linear; these are leucine, isoleucine, and valine. Proline is the only proteinogenic amino acid whose side-group links to the α-amino group and, thus, is also the only proteinogenic amino acid containing a secondary amine at this position. In chemical terms, proline is, therefore, an imino acid, since it lacks a primary amino group, although it is still classed as an amino acid in the current biochemical nomenclature, and may also be called an "N-alkylated alpha-amino acid".

Zwitterions

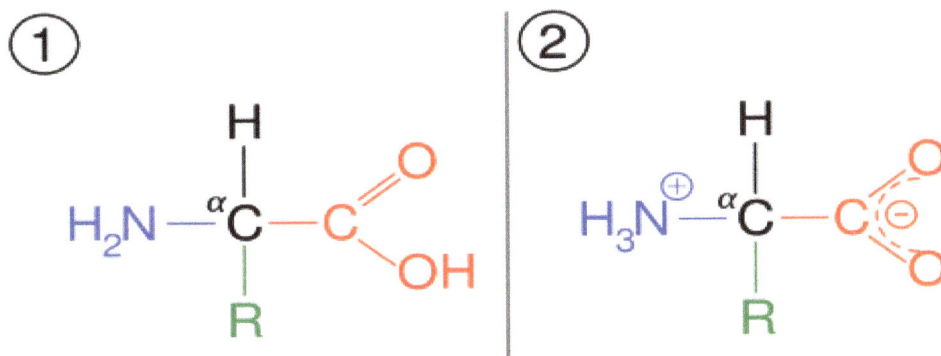

An amino acid in its (1) un-ionized and (2) zwitterionic forms

The α-carboxylic acid group of amino acids is a weak acid, meaning that it releases a hydron (such as a proton) at moderate pH values. In other words, carboxylic acid groups ($-CO_2H$) can be deprotonated to become negative carboxylates ($-CO_2^-$). The negatively charged carboxylate ion predominates at pH values greater than the pKa of the carboxylic acid group (mean for the 20 common amino acids is about 2.2, see the table of amino acid structures above). In a complementary fashion, the α-amine of amino acids is a weak base, meaning that it accepts a proton at moderate pH values. In other words, α-amino groups (NH_2-) can be protonated to become positive α-ammonium groups ($^+NH_3-$). The positively charged α-ammonium group predominates at pH values less than the pKa of the α-ammonium group (mean for the 20 common α-amino acids is about 9.4).

Because all amino acids contain amine and carboxylic acid functional groups, they share amphiprotic properties. Below pH 2.2, the predominant form will have a neutral carboxylic acid group and a positive α-ammonium ion (net charge +1), and above pH 9.4, a negative carboxylate and neutral α-amino group (net charge −1). But at pH between 2.2 and 9.4, an amino acid usually contains both a negative carboxylate and a positive α-ammonium group, as shown in structure (2) on the right, so has net zero charge. This molecular state is known as a zwitterion, from the German Zwitter meaning *hermaphrodite* or *hybrid*. The fully neutral form (structure (1) on the right) is a very minor species in aqueous solution throughout the pH range (less than 1 part in

10^7). Amino acids exist as zwitterions also in the solid phase, and crystallize with salt-like properties unlike typical organic acids or amines.

Isoelectric Point

The variation in titration curves when the amino acids are grouped by category can be seen here. With the exception of tyrosine, using titration to differentiate between hydrophobic amino acids is problematic.

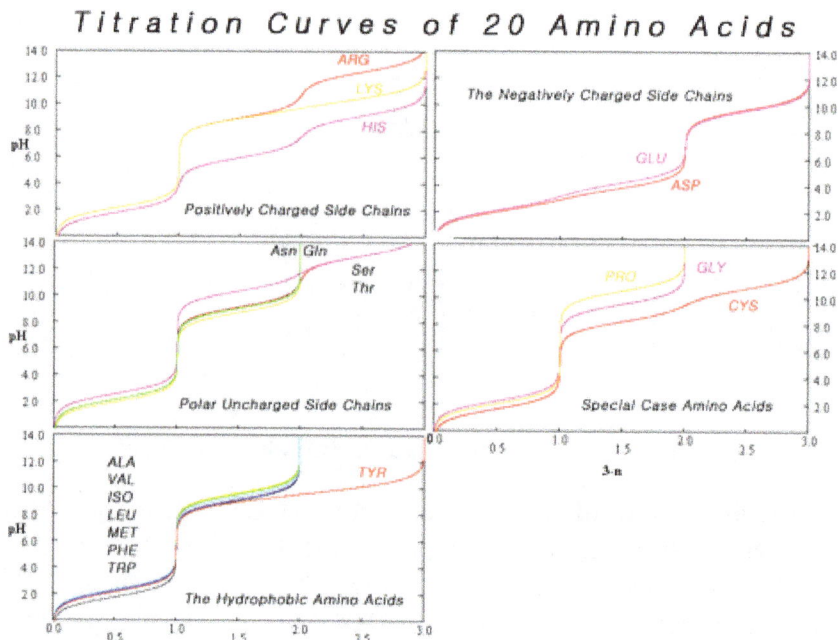

Composite of Titration Curves Grouped by Side Chain Category using applet

At pH values between the two pKa values, the zwitterion predominates, but coexists in dynamic equilibrium with small amounts of net negative and net positive ions. At the exact midpoint between the two pKa values, the trace amount of net negative and trace of net positive ions exactly balance, so that average net charge of all forms present is zero. This pH is known as the isoelectric point pI, so pI = ½(pKa$_1$ + pKa$_2$). The individual amino acids all have slightly different pKa values, so have different isoelectric points. For amino acids with charged side-chains, the pKa of the side-chain is involved. Thus for Asp, Glu with negative side-chains, pI = ½(pKa$_1$ + pKa$_R$), where pKa$_R$ is the side-chain pKa. Cysteine also has potentially negative side-chain with pKa$_R$ = 8.14, so pI should be calculated as for Asp and Glu, even though the side-chain is not significantly charged at neutral pH. For His, Lys, and Arg with positive side-chains, pI = ½(pKa$_R$ + pKa$_2$). Amino acids have zero mobility in electrophoresis at their isoelectric point, although this behaviour is more usually exploited for peptides and proteins than single amino acids. Zwitterions have minimum solubility at their isoelectric point and some amino acids (in particular, with non-polar side-chains) can be isolated by precipitation from water by adjusting the pH to the required isoelectric point.

Occurrence and Functions in Biochemistry

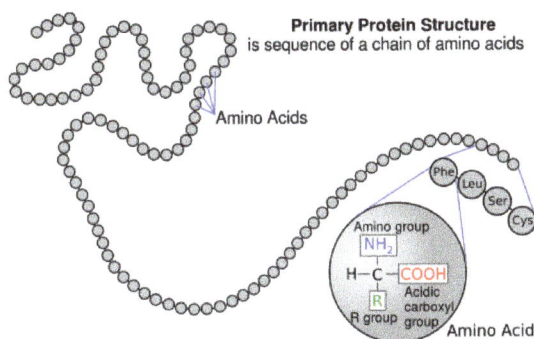

Primary Protein Structure is sequence of a chain of amino acids

A polypeptide is an unbranched chain of amino acids.

Proteinogenic Amino Acids

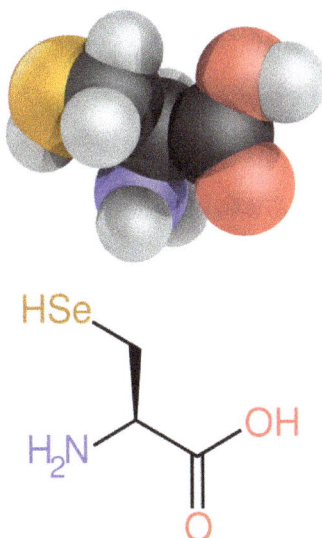

The amino acid selenocysteine

Amino acids are the structural units (monomers) that make up proteins. They join together to form short polymer chains called peptides or longer chains called either polypeptides or proteins. These polymers are linear and unbranched, with each amino acid within the chain attached to two neighboring amino acids. The process of making proteins is called *translation* and involves the step-by-step addition of amino acids to a growing protein chain by a ribozyme that is called a ribosome. The order in which the amino acids are added is read through the genetic code from an mRNA template, which is a RNA copy of one of the organism's genes.

Twenty-two amino acids are naturally incorporated into polypeptides and are called proteinogenic or natural amino acids. Of these, 20 are encoded by the universal genetic code. The remaining 2, selenocysteine and pyrrolysine, are incorporated into proteins by unique synthetic mechanisms. Selenocysteine is incorporated when the mRNA being translated includes a SECIS element, which causes the UGA codon to encode selenocysteine instead of a stop codon. Pyrrolysine is used by some methanogenic archaea in enzymes that they use to produce methane. It is coded for with the codon UAG, which is normally a stop codon in other organisms. This UAG codon is followed by a PYLIS downstream sequence.

Non-proteinogenic Amino Acids

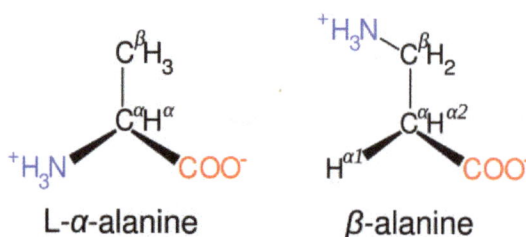

β-alanine and its α-alanine isomer

Aside from the 22 proteinogenic amino acids, there are many other amino acids that are called *non-proteinogenic*. Those either are not found in proteins (for example carnitine, GABA) or are not produced directly and in isolation by standard cellular machinery (for example, hydroxyproline and selenomethionine).

Non-proteinogenic amino acids that are found in proteins are formed by post-translational modification, which is modification after translation during protein synthesis. These modifications are often essential for the function or regulation of a protein; for example, the carboxylation of glutamate allows for better binding of calcium cations, and the hydroxylation of proline is critical for maintaining connective tissues. Another example is the formation of hypusine in the translation initiation factor EIF5A, through modification of a lysine residue. Such modifications can also determine the localization of the protein, e.g., the addition of long hydrophobic groups can cause a protein to bind to a phospholipid membrane.

Some non-proteinogenic amino acids are not found in proteins. Examples include lanthionine, 2-aminoisobutyric acid, dehydroalanine, and the neurotransmitter gamma-aminobutyric acid. Non-proteinogenic amino acids often occur as intermediates in the metabolic pathways for standard amino acids – for example, ornithine and citrulline occur in the urea cycle, part of amino acid catabolism. A rare exception to the dominance of α-amino acids in biology is the β-amino acid beta alanine (3-aminopropanoic acid), which is used in plants and microorganisms in the synthesis of pantothenic acid (vitamin B_5), a component of coenzyme A.

D-amino Acid Natural Abundance

D-isomers are uncommon in live organisms. For instance, gramicidin is a polypeptide made up from mixture of D- and L-amino acids. Other compounds containing D-amino acids are tyrocidine and valinomycin. These compounds disrupt bacterial cell walls, particularly in Gram-positive bacteria. Only 837 D-amino acids were found in Swiss-Prot database (187 million amino acids analysed).

Non-standard Amino Acids

The 20 amino acids that are encoded directly by the codons of the universal genetic code are called *standard* or *canonical* amino acids. The others are called *non-standard* or *non-canonical*. Most of the non-standard amino acids are also non-proteinogenic (i.e. they cannot be used to build proteins), but three of them are proteinogenic, as they can be used to build proteins by exploiting information not encoded in the universal genetic code.

The three non-standard proteinogenic amino acids are selenocysteine (present in many non-eukaryotes as well as most eukaryotes, but not coded directly by DNA), pyrrolysine (found only in some archaea and one bacterium), and *N*-formylmethionine (which is often the initial amino acid of proteins in bacteria, mitochondria, and chloroplasts). For example, 25 human proteins include selenocysteine (Sec) in their primary structure, and the structurally characterized enzymes (selenoenzymes) employ Sec as the catalytic moiety in their active sites. Pyrrolysine and selenocysteine are encoded via variant codons. For example, selenocysteine is encoded by stop codon and SECIS element.

In Human Nutrition

When taken up into the human body from the diet, the 20 standard amino acids either are used to synthesize proteins and other biomolecules or are oxidized to urea and carbon dioxide as a source of energy. The oxidation pathway starts with the removal of the amino group by a transaminase; the amino group is then fed into the urea cycle. The other product of transamidation is a keto acid that enters the citric acid cycle. Glucogenic amino acids can also be converted into glucose, through gluconeogenesis. Of the 20 standard amino acids, nine (His, Ile, Leu, Lys, Met, Phe, Thr, Trp and Val), are called essential amino acids because the human body cannot synthesize them from other compounds at the level needed for normal growth, so they must be obtained from food. In addition, cysteine, taurine, tyrosine, and arginine are considered semiessential amino-acids in children (though taurine is not technically an amino acid), because the metabolic pathways that synthesize these amino acids are not fully developed. The amounts required also depend on the age and health of the individual, so it is hard to make general statements about the dietary requirement for some amino acids. Dietary exposure to the non-standard amino acid BMAA has been linked to human neurodegenerative diseases, including ALS.

Non-protein Functions

Human Biosynthesis Pathway for Trace Amines and Catecholamines

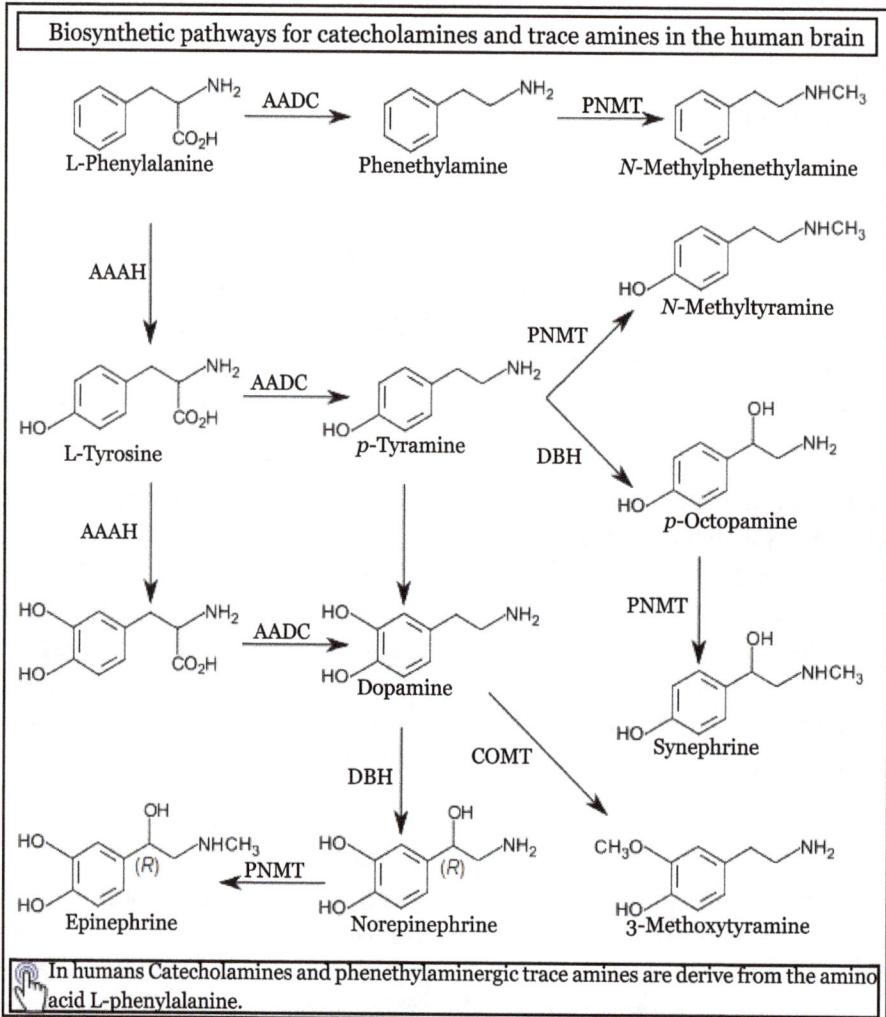

Biosynthetic pathways for catecholamines and trace amines in the human brain

In humans Catecholamines and phenethylaminergic trace amines are derive from the amino acid L-phenylalanine.

Catecholamines and trace amines are synthesized from phenylalanine and tyrosine in humans.

In humans, non-protein amino acids also have important roles as metabolic intermediates, such as in the biosynthesis of the neurotransmitter gamma-amino-butyric acid (GABA). Many amino acids are used to synthesize other molecules, for example:

- Tryptophan is a precursor of the neurotransmitter serotonin.

- Tyrosine (and its precursor phenylalanine) are precursors of the catecholamine neurotransmitters dopamine, epinephrine and norepinephrine and various trace amines.

- Phenylalanine is a precursor of phenethylamine and tyrosine in humans. In plants, it is a precursor of various phenylpropanoids, which are important in plant metabolism.

- Glycine is a precursor of porphyrins such as heme.

- Arginine is a precursor of nitric oxide.

- Ornithine and S-adenosylmethionine are precursors of polyamines.

- Aspartate, glycine, and glutamine are precursors of nucleotides.

However, not all of the functions of other abundant non-standard amino acids are known.

Some non-standard amino acids are used as defenses against herbivores in plants. For example, canavanine is an analogue of arginine that is found in many legumes, and in particularly large amounts in *Canavalia gladiata* (sword bean). This amino acid protects the plants from predators such as insects and can cause illness in people if some types of legumes are eaten without processing. The non-protein amino acid mimosine is found in other species of legume, in particular *Leucaena leucocephala*. This compound is an analogue of tyrosine and can poison animals that graze on these plants.

Uses in Industry

Amino acids are used for a variety of applications in industry, but their main use is as additives to animal feed. This is necessary, since many of the bulk components of these feeds, such as soybeans, either have low levels or lack some of the essential amino acids: lysine, methionine, threonine, and tryptophan are most important in the production of these feeds. In this industry, amino acids are also used to chelate metal cations in order to improve the absorption of minerals from supplements, which may be required to improve the health or production of these animals.

The food industry is also a major consumer of amino acids, in particular, glutamic acid, which is used as a flavor enhancer, and aspartame (aspartyl-phenylalanine-1-methyl ester) as a low-calorie artificial sweetener. Similar technology to that used for animal nutrition is employed in the human nutrition industry to alleviate symptoms of mineral deficiencies, such as anemia, by improving mineral absorption and reducing negative side effects from inorganic mineral supplementation.

The chelating ability of amino acids has been used in fertilizers for agriculture to facilitate the delivery of minerals to plants in order to correct mineral deficiencies, such as iron chlorosis. These fertilizers are also used to prevent deficiencies from occurring and improving the overall health of the plants. The remaining production of amino acids is used in the synthesis of drugs and cosmetics.

Similarly, some amino acids derivatives are used in pharmaceutical industry. They include 5-HTP (5-hydroxytryptophan) used for experimental treatment of depression,

L-DOPA (L-dihydroxyphenylalanine) for Parkinson's treatment, and eflornithine drug that inhibits ornithine decarboxylase and used in the treatment of sleeping sickness.

Expanded Genetic Code

Since 2001, 40 non-natural amino acids have been added into protein by creating a unique codon (recoding) and a corresponding transfer-RNA:aminoacyl – tRNA-synthetase pair to encode it with diverse physicochemical and biological properties in order to be used as a tool to exploring protein structure and function or to create novel or enhanced proteins.

Nullomers

Nullomers are codons that in theory code for an amino acid, however in nature there is a selective bias against using this codon in favor of another, for example bacteria prefer to use CGA instead of AGA to code for arginine. This creates some sequences that do not appear in the genome. This characteristic can be taken advantage of and used to create new selective cancer-fighting drugs and to prevent cross-contamination of DNA samples from crime-scene investigations.

Chemical Building Blocks

Amino acids are important as low-cost feedstocks. These compounds are used in chiral pool synthesis as enantiomerically pure building-blocks.

Amino acids have been investigated as precursors chiral catalysts, e.g., for asymmetric hydrogenation reactions, although no commercial applications exist.

Biodegradable Plastics

Amino acids are under development as components of a range of biodegradable polymers. These materials have applications as environmentally friendly packaging and in medicine in drug delivery and the construction of prosthetic implants. These polymers include poly-peptides, polyamides, polyesters, polysulfides, and polyurethanes with amino acids either forming part of their main chains or bonded as side-chains. These modifications alter the physical properties and reactivities of the polymers. An interesting example of such mate-rials is polyaspartate, a water-soluble biodegradable polymer that may have applications in disposable diapers and agriculture. Due to its solubility and ability to chelate metal ions, polyaspartate is also being used as a biodegradeable anti-scaling agent and a corrosion inhibitor. In addition, the aromatic amino acid tyrosine is being developed as a possible replacement for toxic phenols such as bisphenol A in the manufacture of polycarbonates.

Reactions

As amino acids have both a primary amine group and a primary carboxyl group, these chemicals can undergo most of the reactions associated with these functional groups.

These include nucleophilic addition, amide bond formation, and imine formation for the amine group, and esterification, amide bond formation, and decarboxylation for the carboxylic acid group. The combination of these functional groups allow amino acids to be effective polydentate ligands for metal-amino acid chelates. The multiple side-chains of amino acids can also undergo chemical reactions. The types of these reactions are determined by the groups on these side-chains and are, therefore, different between the various types of amino acid.

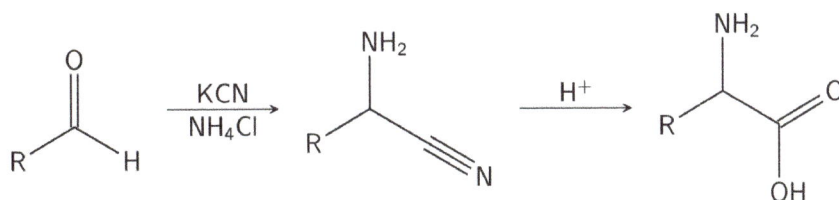

The Strecker amino acid synthesis

Chemical Synthesis

Several methods exist to synthesize amino acids. One of the oldest methods begins with the bromination at the α-carbon of a carboxylic acid. Nucleophilic substitution with ammonia then converts the alkyl bromide to the amino acid. In alternative fashion, the Strecker amino acid synthesis involves the treatment of an aldehyde with potassium cyanide and ammonia, this produces an α-amino nitrile as an intermediate. Hydrolysis of the nitrile in acid then yields a α-amino acid. Using ammonia or ammonium salts in this reaction gives unsubstituted amino acids, whereas substituting primary and secondary amines will yield substituted amino acids. Likewise, using ketones, instead of aldehydes, gives α,α-disubstituted amino acids. The classical synthesis gives racemic mixtures of α-amino acids as products, but several alternative procedures using asymmetric auxiliaries or asymmetric catalysts have been developed.

At the current time, the most-adopted method is an automated synthesis on a solid support (e.g., polystyrene beads), using protecting groups (e.g., Fmoc and t-Boc) and activating groups (e.g., DCC and DIC).

Peptide Bond Formation

As both the amine and carboxylic acid groups of amino acids can react to form amide bonds, one amino acid molecule can react with another and become joined through an amide linkage. This polymerization of amino acids is what creates proteins. This condensation reaction yields the newly formed peptide bond and a molecule of water. In cells, this reaction does not occur directly; instead, the amino acid is first activated by attachment to a transfer RNA molecule through an ester bond. This aminoacyl-tRNA is produced in an ATP-dependent reaction carried out by an aminoacyl tRNA synthetase. This aminoacyl-tRNA is then a substrate for the ribosome, which catalyzes the attack of the amino group of the elongating protein chain on the ester bond. As a result of this mechanism, all proteins made by ribosomes are synthesized starting at their N-terminus and moving toward their C-terminus.

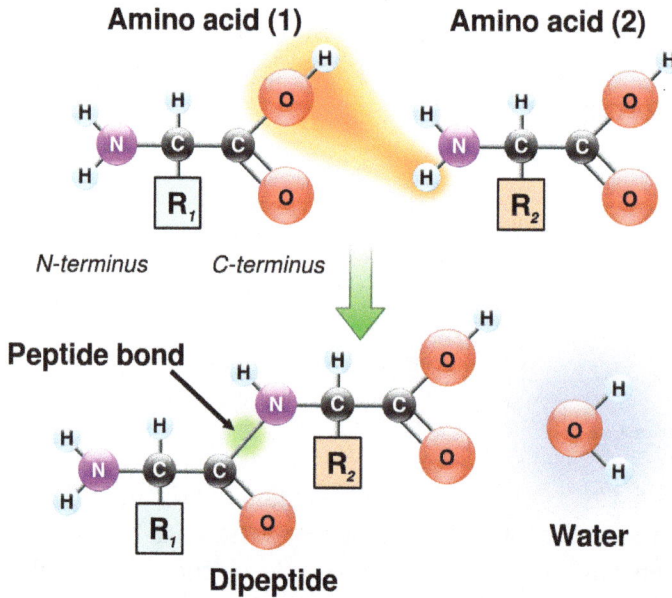

The condensation of two amino acids to form a *dipeptide* through a *peptide bond*

However, not all peptide bonds are formed in this way. In a few cases, peptides are synthesized by specific enzymes. For example, the tripeptide glutathione is an essential part of the defenses of cells against oxidative stress. This peptide is synthesized in two steps from free amino acids. In the first step, gamma-glutamylcysteine synthetase condenses cysteine and glutamic acid through a peptide bond formed between the side-chain carboxyl of the glutamate (the gamma carbon of this side-chain) and the amino group of the cysteine. This dipeptide is then condensed with glycine by glutathione synthetase to form glutathione.

In chemistry, peptides are synthesized by a variety of reactions. One of the most-used in solid-phase peptide synthesis uses the aromatic oxime derivatives of amino acids as activated units. These are added in sequence onto the growing peptide chain, which is attached to a solid resin support. The ability to easily synthesize vast numbers of different peptides by varying the types and order of amino acids (using combinatorial chemistry) has made peptide synthesis particularly important in creating libraries of peptides for use in drug discovery through high-throughput screening.

Biosynthesis

In plants, nitrogen is first assimilated into organic compounds in the form of glutamate, formed from alpha-ketoglutarate and ammonia in the mitochondrion. In order to form other amino acids, the plant uses transaminases to move the amino group to another alpha-keto carboxylic acid. For example, aspartate aminotransferase converts glutamate and oxaloacetate to alpha-ketoglutarate and aspartate. Other organisms use transaminases for amino acid synthesis, too.

Nonstandard amino acids are usually formed through modifications to standard amino acids. For example, homocysteine is formed through the transsulfuration pathway or by the demethylation of methionine via the intermediate metabolite S-adenosyl methionine, while hydroxyproline is made by a posttranslational modification of proline.

Microorganisms and plants can synthesize many uncommon amino acids. For example, some microbes make 2-aminoisobutyric acid and lanthionine, which is a sulfide-bridged derivative of alanine. Both of these amino acids are found in peptidic lantibiotics such as alamethicin. However, in plants, 1-aminocyclopropane-1-carboxylic acid is a small disubstituted cyclic amino acid that is a key intermediate in the production of the plant hormone ethylene.

Catabolism

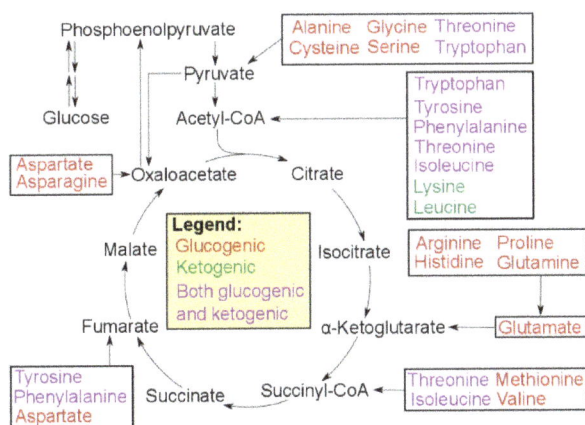

Catabolism of proteinogenic amino acids. Amino acids can be classified according to the properties of their main products as either of the following:
* *Glucogenic*, with the products having the ability to form glucose by gluconeogenesis
* *Ketogenic*, with the products not having the ability to form glucose. These products may still be used for ketogenesis or lipid synthesis.
* Amino acids catabolized into both glucogenic and ketogenic products.

Amino acids must first pass out of organelles and cells into blood circulation via amino acid transporters, since the amine and carboxylic acid groups are typically ionized. Degradation of an amino acid, occurring in the liver and kidneys, often involves deamination by moving its amino group to alpha-ketoglutarate, forming glutamate. This process involves transaminases, often the same as those used in amination during synthesis. In many vertebrates, the amino group is then removed through the urea cycle and is excreted in the form of urea. However, amino acid degradation can produce uric acid or ammonia instead. For example, serine dehydratase converts serine to pyruvate and ammonia. After removal of one or more amino groups, the remainder of the molecule can sometimes be used to synthesize new amino acids, or it can be used for energy by entering glycolysis or the citric acid cycle, as detailed in image at right.

Physicochemical Properties of Amino Acids

The 20 amino acids encoded directly by the genetic code can be divided into several groups based on their properties. Important factors are charge, hydrophilicity or hydrophobicity, size, and functional groups. These properties are important for protein structure and protein–protein interactions. The water-soluble proteins tend to have their hydrophobic residues (Leu, Ile, Val, Phe, and Trp) buried in the middle of the protein, whereas hydrophilic side-chains are exposed to the aqueous solvent. (Note that in biochemistry, a residue refers to a specific monomer within the polymeric chain of a polysaccharide, protein or nucleic acid.) The integral membrane proteins tend to have outer rings of exposed hydrophobic amino acids that anchor them into the lipid bilayer. In the case part-way between these two extremes, some peripheral membrane proteins have a patch of hydrophobic amino acids on their surface that locks onto the membrane. In similar fashion, proteins that have to bind to positively charged molecules have surfaces rich with negatively charged amino acids like glutamate and aspartate, while proteins binding to negatively charged molecules have surfaces rich with positively charged chains like lysine and arginine. There are different hydrophobicity scales of amino acid residues.

Some amino acids have special properties such as cysteine, that can form covalent disulfide bonds to other cysteine residues, proline that forms a cycle to the polypeptide backbone, and glycine that is more flexible than other amino acids.

Many proteins undergo a range of posttranslational modifications, when additional chemical groups are attached to the amino acids in proteins. Some modifications can produce hydrophobic lipoproteins, or hydrophilic glycoproteins. These type of modification allow the reversible targeting of a protein to a membrane. For example, the addition and removal of the fatty acid palmitic acid to cysteine residues in some signaling proteins causes the proteins to attach and then detach from cell membranes.

Table of Standard Amino Acid Abbreviations and Properties

Amino Acid	3-Letter	1-Letter	Side-chain class	Side-chain polarity	Side-chain charge (pH 7.4)	Hydropathy index	Absorbance λ_{max}(nm)	ε at λ_{max} (mM^{-1} cm^{-1})	MW (Weight)
Alanine	Ala	A	aliphatic	nonpolar	neutral	1.8			89.094
Arginine	Arg	R	basic	basic polar	positive	−4.5			174.203
Asparagine	Asn	N	acid (amide)	polar	neutral	−3.5			132.119

Amino Acid	3-Letter	1-Letter	Side-chain class	Side-chain polarity	Side-chain charge (pH 7.4)	Hydropathy index	Absorbance λ_{max}(nm)	ε at λ_{max} (mM^{-1} cm^{-1})	MW (Weight)
Aspartic acid	Asp	D	acid (amide)	acidic polar	negative	−3.5			133.104
Cysteine	Cys	C	sulfur-containing	nonpolar	neutral	2.5	250	0.3	121.154
Glutamic acid	Glu	E	acid (amide)	acidic polar	negative	−3.5			147.131
Glutamine	Gln	Q	acid (amide)	polar	neutral	−3.5			146.146
Glycine	Gly	G	aliphatic	nonpolar	neutral	−0.4			75.067
Histidine	His	H	basic	basic polar	positive(10%) neutral(90%)	−3.2	211	5.9	155.156
Isoleucine	Ile	I	aliphatic	nonpolar	neutral	4.5			131.175
Leucine	Leu	L	aliphatic	nonpolar	neutral	3.8			131.175
Lysine	Lys	K	basic	basic polar	positive	−3.9			146.189
Methionine	Met	M	sulfur-containing	nonpolar	neutral	1.9			149.208
Phenylalanine	Phe	F	aromatic	nonpolar	neutral	2.8	257, 206, 188	0.2, 9.3, 60.0	165.192
Proline	Pro	P	cyclic	nonpolar	neutral	−1.6			115.132
Serine	Ser	S	hydroxyl-containing	polar	neutral	−0.8			105.093
Threonine	Thr	T	hydroxyl-containing	polar	neutral	−0.7			119.12
Tryptophan	Trp	W	aromatic	nonpolar	neutral	−0.9	280, 219	5.6, 47.0	204.228
Tyrosine	Tyr	Y	aromatic	polar	neutral	−1.3	274, 222, 193	1.4, 8.0, 48.0	181.191
Valine	Val	V	aliphatic	nonpolar	neutral	4.2			117.148

Two additional amino acids are in some species coded for by codons that are usually interpreted as stop codons:

21st and 22nd amino acids	3-Letter	1-Letter	MW(Weight)
Selenocysteine	Sec	U	159.065
Pyrrolysine	Pyl	O	273.325

In addition to the specific amino acid codes, placeholders are used in cases where chemical or crystallographic analysis of a peptide or protein cannot conclusively determine the identity of a residue. They are also used to summarise conserved protein sequence motifs. The use of single letters to indicate sets of similar residues is similar to the use of abbreviation codes for degenerate bases.

Ambiguous Amino Acids	3-Letter	1-Letter	
Any / unknown	Xaa	X	All
Asparagine or aspartic acid	Asx	B	D, N
Glutamine or glutamic acid	Glx	Z	E, Q
Leucine or Isoleucine	Xle	J	I, L
Hydrophobic		Φ	V, I, L, F, W, Y, M
Aromatic		Ω	F, W, Y
Aliphatic		Ψ	V, I, L, M
Small		π	P, G, A, S
Hydrophilic		ζ	S, T, H, N, Q, E, D, K, R
Positively charged		+	K, R
Negatively charged		–	D, E

Unk is sometimes used instead of Xaa, but is less standard.

In addition, many non-standard amino acids have a specific code. For example, several peptide drugs, such as Bortezomib and MG132, are artificially synthesized and retain their protecting groups, which have specific codes. Bortezomib is Pyz-Phe-boroLeu, and MG132 is Z-Leu-Leu-Leu-al. To aid in the analysis of protein structure, photo-reactive amino acid analogs are available. These include photoleucine (pLeu) and photomethionine (pMet).

Amine

Primary amine	Secondary amine	Tertiary amine

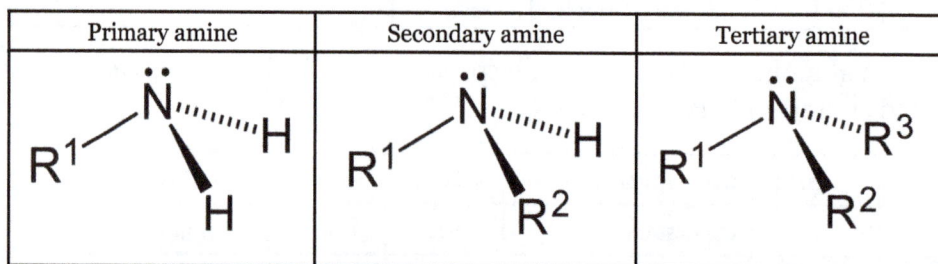

In organic chemistry, amines are compounds and functional groups that contain a basic nitrogen atom with a lone pair. Amines are formally derivatives of ammonia, where in one or more hydrogen atoms have been replaced by a substituent such as an alkyl or aryl group. (These may respectively be called alkylamines and arylamines; amines in which both types of substituent are attached to one nitrogen atom may be called alkylarylamines.) Important amines include amino acids, biogenic amines, trimethylamine, and aniline. Inorganic de-rivatives of ammonia are also called amines, such as chloramine ($NClH_2$)

Compounds with a nitrogen atom attached to a carbonyl group, thus having the structure R–CO–NR′R″, are called amides and have different chemical properties from amines.

Classes of Amines

An aliphatic amine has no aromatic ring attached directly to the nitrogen atom. Aromatic amines have the nitrogen atom connected to an aromatic ring as in the various anilines. The aromatic ring decreases the alkalinity of the amine, depending on its substituents. The presence of an amine group strongly increases the reactivity of the aromatic ring, due to an electron-donating effect.

Amines are organized into four subcategories:

- Primary amines — Primary amines arise when one of three hydrogen atoms in ammonia is replaced by an alkyl or aromatic. Important primary alkyl amines include methylamine, ethanolamine (2-aminoethanol), and the buffering agent tris, while primary aromatic amines include aniline.

- Secondary amines — Secondary amines have two organic substituents (alkyl, aryl or both) bound to N together with one hydrogen (or no hydrogen if one of the substituent bonds is double). Important representatives include dimethylamine and methylethanolamine, while an example of an aromatic amine would be diphenylamine.

- Tertiary amines — In tertiary amines, all three hydrogen atoms are replaced by organic substituents. Examples include trimethylamine, which has a distinctively fishy smell, or triphenylamine.

- Cyclic amines — Cyclic amines are either secondary or tertiary amines. Examples of cyclic amines include the 3-membered ring aziridine and the six-membered ring piperidine. N-methylpiperidine and N-phenylpiperidine are examples of cyclic tertiary amines.

It is also possible to have four organic substituents on the nitrogen. These species are not amines but are quaternary ammonium cations and have a charged nitrogen center. Quaternary ammonium salts exist with many kinds of anions.

Naming Conventions

Amines are named in several ways. Typically, the compound is given the prefix "amino-" or the suffix: "-amine". The prefix "N-" shows substitution on the nitrogen atom. An organic compound with multiple amino groups is called a diamine, triamine, tetraamine and so forth.

Systematic names for some common amines:

Lower amines are named with the suffix -amine.	Higher amines have the prefix amino as a functional group. IUPAC however does not recommend this convention, but prefers the alkanamine form, e.g. pentan-2-amine.
methylamine	2-aminopentane (or sometimes: pent-2-yl-amine or pentan-2-amine)

Physical properties

Hydrogen bonding significantly influences the properties of primary and secondary amines. Thus the boiling point of amines is higher than those of the corresponding phosphines, but generally lower than those of the corresponding alcohols. For example, methyl and ethyl amines are gases under standard conditions, whereas the corresponding methyl and ethyl alcohols are liquids. Gaseous amines possess a characteristic ammonia smell, liquid amines have a distinctive "fishy" smell.

Also reflecting their ability to form hydrogen bonds, most aliphatic amines display some solubility in water. Solubility decreases with the increase in the number of carbon atoms. Aliphatic amines display significant solubility in organic solvents, especially polar organic solvents. Primary amines react with ketones such as acetone.

The aromatic amines, such as aniline, have their lone pair electrons conjugated into the benzene ring, thus their tendency to engage in hydrogen bonding is diminished. Their boiling points are high and their solubility in water is low.

Chirality

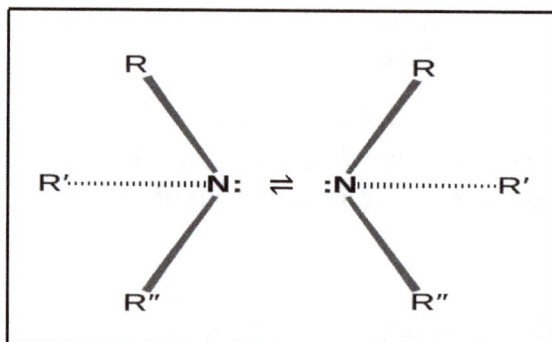

Inversion of an amine. The pair of dots represents the lone electron pair on the nitrogen atom.

Amines of the type NHRR' and NRR'R" are chiral: the nitrogen atom bears four substituents counting the lone pair. The energy barrier for the inversion of the stereocenter is relatively low, e.g., ~7 kcal/mol for a trialkylamine. The interconversion of the stereoisomers has been compared to the inversion of an open umbrella into a strong wind. Because of this low barrier, amines such as NHRR' cannot be resolved optically and NRR'R" can only be resolved when the R, R', and R" groups are constrained in cyclic structures such as aziridines. Quaternary ammonium salts with four distinct groups on the nitrogen are capable of exhibiting optical activity.

Properties as Bases

Like ammonia, amines are bases. Compared to alkali metal hydroxides, amines are weaker. The basicity of amines depends on:

1. The electronic properties of the substituents (alkyl groups enhance the basicity, aryl groups diminish it).

2. Steric hindrance offered by the groups on nitrogen.

3. The degree of solvation of the protonated amine.

The nitrogen atom features a lone electron pair that can bind H^+ to form an ammonium ion R_3NH^+. The lone electron pair is represented in this article by a two dots above or next to the N. The water solubility of simple amines is largely due to hydrogen bonding between protons in the water molecules and these lone electron pairs.

- Inductive effect of alkyl groups

Ions of compound	K_b
Ammonia NH_3	1.8×10^{-5} M
Propylamine $CH_3CH_2CH_2NH_2$	4.7×10^{-4} M
Isopropylamine $(CH_3)_2CHNH_2$	3.4×10^{-4} M
Methylamine CH_3NH_2	4.4×10^{-4} M
Dimethylamine $(CH_3)_2NH$	5.4×10^{-4} M
Trimethylamine $(CH_3)_3N$	5.9×10^{-5} M

The +I effect of alkyl groups raises the energy of the lone pair of electrons, thus elevating the basicity. Thus the basicity of an amine may be expected to increase with the number of alkyl groups on the amine. However, there is no strict trend in this regard, as basicity is also governed by other factors mentioned above. Consider the K_b values of the methyl amines given above. The increase in K_b from methylamine to dimethylamine may be attributed to the +I effect; however, there is a decrease from dimethylamine to trimethyl amine due to the predominance of steric hindrance offered by the three methyl groups to the approaching Brönsted acid.

- Mesomeric effect of aromatic systems

Ions of compound	K_b
Ammonia NH_3	1.8×10^{-5} M
Aniline $C_6H_5NH_2$	3.8×10^{-10} M
4-Methylaniline $4\text{-}CH_3C_6H_4NH_2$	1.2×10^{-9} M
2-Nitroaniline	1.5×10^{-15} M
3-Nitroaniline	2.8×10^{-13} M
4-Nitroaniline	9.5×10^{-14} M

The −M effect of aromatic ring delocalises the lone pair of electrons on nitrogen into the ring, resulting in decreased basicity. Substituents on the aromatic ring, and their positions relative to the amine group may also considerably alter basicity as seen above.

The solvation of protonated amines changes upon their conversion to ammonium compounds. Typically salts of ammonium compounds exhibit the following order of solubility in water: primary ammonium ($RNH+3$) > secondary ammonium ($R2NH+2$) > tertiary ammonium (R_3NH^+). Quaternary ammonium salts usually exhibit the lowest solubility of the series.

In sterically hindered amines, as in the case of trimethylamine, the protonated form is not well-solvated. For this reason the parent amine is less basic than expected. In the case of aprotic polar solvents (like DMSO and DMF), wherein the extent of solvation is not as high as in protic polar solvents (like water and methanol), the basicity of amines is almost solely governed by the electronic factors within the molecule.

Synthesis

Alkylation

The most industrially significant amines are prepared from ammonia by alkylation with alcohols:

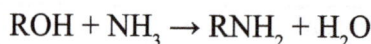

$$ROH + NH_3 \rightarrow RNH_2 + H_2O$$

These reactions require catalysts, specialized apparatus, and additional purification measures since the selectivity can be problematic. The same amines can be prepared by treatment of haloalkanes with ammonia and amines:

$$RX + 2\ R'NH_2 \rightarrow RR'NH + [RR'NH_2]X$$

Such reactions, which are most useful for alkyl iodides and bromides, are rarely employed because the degree of alkylation is difficult to control. Selectivity can be improved via the delepine reaction, although this is rarely employed on an industrial scale.

Reductive Routes

Via the process of hydrogenation, nitriles are reduced to amines using hydrogen in the presence of a nickel catalyst. Reactions are sensitive to acidic or alkaline conditions, which can cause hydrolysis of the −CN group. $LiAlH_4$ is more commonly employed for the reduction of nitriles on the laboratory scale. Similarly, $LiAlH_4$ reduces amides to amines. Many amines are produced from aldehydes and ketones via reductive amination, which can either proceed catalytically or stoichiometrically.

Aniline ($C_6H_5NH_2$) and its derivatives are prepared by reduction of the nitroaromatics. In industry, hydrogen is the preferred reductant, whereas in the laboratory, tin and iron are often employed.

Specialized Methods

Many laboratory methods exist for the preparation of amines, many of these methods being rather specialized.

Reaction name	Substrate	Comment
Gabriel synthesis	Organohalide	Reagent: potassium phthalimide
Staudinger reduction	Azide	This reaction also takes place with a reducing agent such as lithium aluminium hydride.
Schmidt reaction	Carboxylic acid	
Aza-Baylis−Hillman reaction	Imine	Synthesis of allylic amines
Birch reduction	Imine	Useful for reactions that trap unstable imine intermediates, such as Grignard reactions with nitriles.
Hofmann degradation	Amide	This reaction is valid for preparation of primary amines only. Gives good yields of primary amines uncontaminated with other amines.
Hofmann elimination	Quaternary ammonium salt	Upon treatment with strong base
Amide reduction	amide	
Nitrile reduction	Nitriles	Either accomplished with reducing agents or by electrosynthesis
Reduction of nitro compounds	Nitro compounds	Can be accomplished with elemental zinc, tin or iron with an acid.
Amine alkylation	Haloalkane	
Delepine reaction	Organohalide	reagent Hexamine
Buchwald−Hartwig reaction	Aryl halide	Specific for aryl amines
Menshutkin reaction	Tertiary amine	Reaction product a quaternary ammonium cation

Reaction name	Substrate	Comment
Hydroamination	Alkenes and alkynes	
Oxime reduction	Oximes	
Leuckart reaction	Ketones and aldehydes	Reductive amination with formic acid and ammonia via an imine intermediate
Hofmann–Löffler reaction	Haloamine	
Eschweiler–Clarke reaction	Amine	Reductive amination with formic acid and formaldehyde via an imine intermediate

Reactions

Alkylation, Acylation, and Sulfonation

Aside from their basicity, the dominant reactivity of amines is their nucleophilicity. Most primary amines are good ligands for metal ions to give coordination complexes. Amines are alkylated by alkyl halides. Acyl chlorides and acid anhydrides react with primary and secondary amines to form amides (the "Schotten–Baumann reaction").

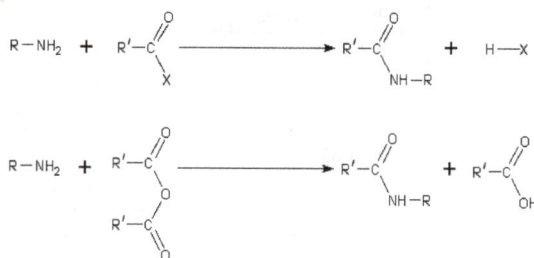

Similarly, with sulfonyl chlorides, one obtains sulfonamides. This transformation, known as the Hinsberg reaction, is a chemical test for the presence of amines. Because amines are basic, they neutralize acids to form the corresponding ammonium salts R_3NH^+. When formed from carboxylic acids and primary and secondary amines, these salts thermally dehydrate to form the corresponding amides.

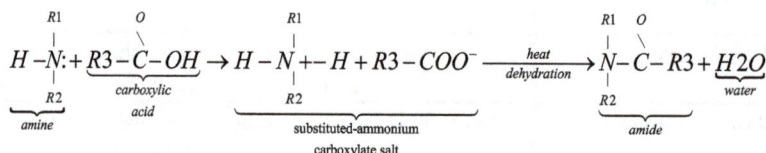

Diazotization

Amines react with nitrous acid to give diazonium salts. The alkyl diazonium salts are of little synthetic importance because they are too unstable. The most important members are derivatives of aromatic amines such as aniline ("phenylamine") (A = aryl or naphthyl):

$$ANH_2 + HNO_2 + HX \rightarrow AN_2^+X^- + 2\,H_2O$$

Anilines and naphthylamines form more stable diazonium salts, which can be isolated in the crystalline form. Diazonium salts undergo a variety of useful transformations involving replacement of the N_2 group with anions. For example, cuprous cyanide gives the corresponding nitriles:

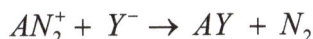

$$AN_2^+ + Y^- \rightarrow AY + N_2$$

Aryldiazonium couple with electron-rich aromatic compounds such as a phenol to form azo compounds. Such reactions are widely applied to the production of dyes.

Conversion to Imines

Imine formation is an important reaction. Primary amines react with ketones and aldehydes to form imines. In the case of formaldehyde (R′ = H), these products typically exist as cyclic trimers.

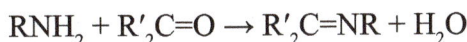

$$RNH_2 + R'_2C{=}O \rightarrow R'_2C{=}NR + H_2O$$

Reduction of these imines gives secondary amines:

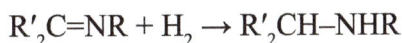

$$R'_2C{=}NR + H_2 \rightarrow R'_2CH{-}NHR$$

Similarly, secondary amines react with ketones and aldehydes to form enamines:

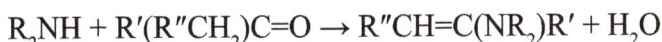

$$R_2NH + R'(R''CH_2)C{=}O \rightarrow R''CH{=}C(NR_2)R' + H_2O$$

Overview

An overview of the reactions of amines is given below:

Reaction name	Reaction product	Comment
Amine alkylation	Amines	Degree of substitution increases
Schotten–Baumann reaction	Amide	Reagents: acyl chlorides, acid anhydrides
Hinsberg reaction	Sulfonamides	Reagents: sulfonyl chlorides
Amine–carbonyl condensation	Imines	
Organic oxidation	Nitroso compounds	Reagent: peroxymonosulfuric acid
Organic oxidation	Diazonium salt	Reagent: nitrous acid
Zincke reaction	Zincke aldehyde	Reagent: pyridinium salts, with primary and secondary amines
Emde degradation	Tertiary amine	Reduction of quaternary ammonium cations

Reaction name	Reaction product	Comment
Hofmann–Martius rearrangement	Aryl-substituted anilines	
Von Braun reaction	Organocyanamide	By cleavage (tertiary amines only) with cyanogen bromide
Hofmann elimination	Alkene	Proceeds by β-elimination of less hindered carbon
Cope reaction	Alkene	Similar to Hofmann elimination
carbylamine reaction	Isonitrile	Primary amines only
Hoffmann's mustard oil test	Isothiocyanate	CS_2 and $HgCl_2$ are used. Thiocyanate smells like mustard.

Biological Activity

Amines are ubiquitous in biology. The breakdown of amino acids releases amines, famously in the case of decaying fish which smell of trimethylamine. Many neurotransmitters are amines, including epinephrine, norepinephrine, dopamine, serotonin, and histamine. Protonated amino groups ($-NH+3$) are the most common positively charged moieties in proteins, specifically in the amino acid lysine. The anionic polymer DNA is typically bound to various amine-rich proteins. Additionally, the terminal charged primary ammonium on lysine forms salt bridges with carboxylate groups of other amino acids in polypeptides, which is one of the primary influences on the three-dimensional structures of proteins.

Application of Amines

Dyes

Primary aromatic amines are used as a starting material for the manufacture of azo dyes. It reacts with nitrous acid to form diazonium salt, which can undergo coupling reaction to form azo compound. As azo-compounds are highly coloured, they are widely used in dyeing industries, such as:

- Methyl Orange
- Direct brown 138
- Sunset yellow FCF
- Ponceau

Drugs

Many drugs are designed to mimic or to interfere with the action of natural amine neurotransmitters, exemplified by the amine drugs:

- Chlorpheniramine is an antihistamine that helps to relieve allergic disorders due to cold, hay fever, itchy skin, insect bites and stings.

- Chlorpromazine is a tranquilizer that sedates without inducing sleep. It is used to relieve anxiety, excitement, restlessness or even mental disorder.

- Ephedrine and phenylephrine, as amine hydrochlorides, are used as decongestants.

- Amphetamine, methamphetamine, and methcathinone are psychostimulant amines that are listed as controlled substances by the US DEA.

- Amitriptyline, imipramine, lofepramine and clomipramine are tricyclic antidepressants and tertiary amines.

- Nortriptyline, desipramine, and amoxapine are tricyclic antidepressants and secondary amines. (The tricyclics are grouped by the nature of the final amine group on the side chain.)

- Substituted tryptamines and phenethylamines are key basic structures for a large variety of psychedelic drugs.

- Opiate analgesics such as morphine, codeine, and heroin are tertiary amines.

Gas Treatment

Aqueous monoethanolamine (MEA), diglycolamine (DGA), diethanolamine (DEA), diisopropanolamine (DIPA) and methyldiethanolamine (MDEA) are widely used industrially for removing carbon dioxide (CO_2) and hydrogen sulfide (H_2S) from natural gas and refinery process streams. They may also be used to remove CO_2 from combustion gases and flue gases and may have potential for abatement of greenhouse gases. Related processes are known as sweetening.

Safety

Low molecular weight amines, such as ethylamine, are toxic, and some are easily absorbed through the skin. Many higher molecular weight amines are, biologically, highly active.

Carboxylic Acid

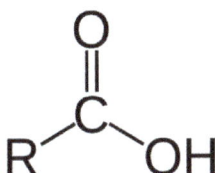

Structure of a carboxylic acid

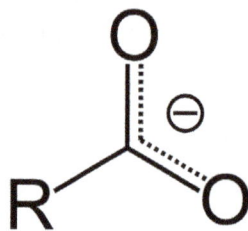

Carboxylate ion

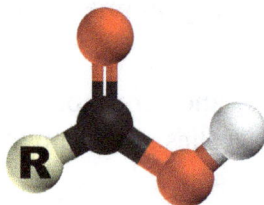

3D structure of a carboxylic acid

A carboxylic acid is an organic compound that contains a carboxyl group (C(O)OH). The general formula of a carboxylic acid is R–C(O)OH, with R referring to the rest of the (possibly quite large) molecule. Carboxylic acids occur widely and include the amino acids (which make up proteins) and acetic acid (which is part of vinegar and occurs in metabolism).

Salts and esters of carboxylic acids are called carboxylates. When a carboxyl group is deprotonated, its conjugate base forms a carboxylate anion. Carboxylate ions are resonance-stabilized, and this increased stability makes carboxylic acids more acidic than alcohols. Carboxylic acids can be seen as reduced or alkylated forms of the Lewis acid carbon dioxide; under some circumstances they can be decarboxylated to yield carbon dioxide.

Example Carboxylic Acids and Nomenclature

Carboxylic acids are commonly named as indicated in the table below. Although rarely used, IUPAC-recommended names also exist. For example, butyric acid ($C_3H_7CO_2H$) is, according to IUPAC guidelines, also known as butanoic acid.

To more easily understand much of the below discussion of reactions involving carboxylic acids it can be helpful to notice that the carboxyl group itself is a "hydroxylated carbonyl group" meaning that two of the carbon atom's four bonds are to an oxygen atom, the carbon atom's third bond is to a second oxygen atom (whose other bond is to a hydrogen atom), and the carbon atom's fourth bond attaches to R. (A carbon atom double bonded to an oxygen atom is a carbonyl group and two of the carbon atom's bonds remain available for bonding. A hydrogen atom bonded to an oxygen atom is a hydroxyl group with the oxygen atom's second bond available for bonding.)

The carboxylate anion R–COO⁻ is usually named with the suffix -*ate*, so acetic acid, for example, becomes acetate ion. In IUPAC nomenclature, carboxylic acids have an -*oic acid* suffix (e.g., octadecanoic acid). For trivial names, the suffix is usually -*ic acid* (e.g., stearic acid).

Straight-chain, saturated carboxylic acids				
Carbon atoms	Common name	IUPAC name	Chemical formula	Common location or use
1	Formic acid	Methanoic acid	HCOOH	Insect stings
2	Acetic acid	Ethanoic acid	CH_3COOH	Vinegar
3	Propionic acid	Propanoic acid	CH_3CH_2COOH	Preservative for stored grains
4	Butyric acid	Butanoic acid	$CH_3(CH_2)_2COOH$	Butter
5	Valeric acid	Pentanoic acid	$CH_3(CH_2)_3COOH$	Valerian
6	Caproic acid	Hexanoic acid	$CH_3(CH_2)_4COOH$	Goat fat
7	Enanthic acid	Heptanoic acid	$CH_3(CH_2)_5COOH$	
8	Caprylic acid	Octanoic acid	$CH_3(CH_2)_6COOH$	Coconuts and breast milk
9	Pelargonic acid	Nonanoic acid	$CH_3(CH_2)_7COOH$	Pelargonium
10	Capric acid	Decanoic acid	$CH_3(CH_2)_8COOH$	Coconut and Palm kernel oil
11	Undecylic acid	Undecanoic acid	$CH_3(CH_2)_9COOH$	
12	Lauric acid	Dodecanoic acid	$CH_3(CH_2)_{10}COOH$	Coconut oil and hand wash soaps
13	Tridecylic acid	Tridecanoic acid	$CH_3(CH_2)_{11}COOH$	
14	Myristic acid	Tetradecanoic acid	$CH_3(CH_2)_{12}COOH$	Nutmeg
15	Pentadecanoic acid		$CH_3(CH_2)_{13}COOH$	
16	Palmitic acid	Hexadecanoic acid	$CH_3(CH_2)_{14}COOH$	Palm oil
17	Margaric acid	Heptadecanoic acid	$CH_3(CH_2)_{15}COOH$	
18	Stearic acid	Octadecanoic acid	$CH_3(CH_2)_{16}COOH$	Chocolate, waxes, soaps, and oils
19	Nonadecylic acid		$CH_3(CH_2)_{17}COOH$	Fats, vegetable oils, pheromone
20	Arachidic acid	Icosanoic acid	$CH_3(CH_2)_{18}COOH$	Peanut oil

Other carboxylic acids	
Compound class	Members
unsaturated monocarboxylic acids	acrylic acid (2-propenoic acid) – CH_2=CHCOOH, used in polymer synthesis
Fatty acids	medium to long-chain saturated and unsaturated monocarboxylic acids, with even number of carbons examples docosahexaenoic acid and eicosapentaenoic acid (nutritional supplements)
Amino acids	the building-blocks of proteins
Keto acids	acids of biochemical significance that contain a ketone group, e.g. acetoacetic acid and pyruvic acid

Aromatic carboxylic acids	benzoic acid, the sodium salt of benzoic acid is used as a food preservative, salicylic acid – a beta hydroxy type found in many skin-care products, phenyl alkanoic acids the class of compounds where a phenyl group is attached to a carboxylic acid.
Dicarboxylic acids	containing two carboxyl groups examples adipic acid the monomer used to produce nylon and aldaric acid – a family of sugar acids
Tricarboxylic acids	containing three carboxyl groups example citric acid – found in citrus fruits and isocitric acid
Alpha hydroxy acids	containing a hydroxy group example glyceric acid, glycolic acid and lactic acid (2-hydroxypropanoic acid) – found in sour milk tartaric acid – found in wine
Divinylether fatty acids	containing a doubly unsaturated carbon chain attached via an ether bond to a fatty acid, found in some plants

Carboxyl Radical

The radical ·COOH (CAS# 2564-86-5) has only a separate fleeting existence. The acid dissociation constant of ·COOH has been measured using electron paramagnetic resonance spectroscopy. The carboxyl group tends to dimerise to form oxalic acid.

Physical Properties

Solubility

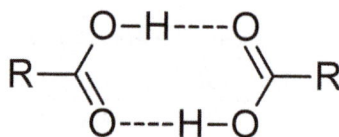

Carboxylic acid dimers

Carboxylic acids are polar. Because they are both hydrogen-bond acceptors (the carbonyl −C=O) and hydrogen-bond donors (the hydroxyl −OH), they also participate in hydrogen bonding. Together the hydroxyl and carbonyl group forms the functional group carboxyl. Carboxylic acids usually exist as dimeric pairs in nonpolar media due to their tendency to "self-associate." Smaller carboxylic acids (1 to 5 carbons) are soluble in water, whereas higher carboxylic acids are less soluble due to the increasing hydrophobic nature of the alkyl chain. These longer chain acids tend to be rather soluble in less-polar solvents such as ethers and alcohols.

Boiling Points

Carboxylic acids tend to have higher boiling points than water, not only because of their increased surface area, but because of their tendency to form stabilised dimers. Carboxylic acids tend to evaporate or boil as these dimers. For boiling to occur, either the dimer bonds must be broken or the entire dimer arrangement must be vaporised, both of which increase the enthalpy of vaporization requirements significantly.

Acidity

Carboxylic acids are Brønsted–Lowry acids because they are proton (H$^+$) donors. They are the most common type of organic acid.

Carboxylic acids are typically weak acids, meaning that they only partially dissociate into H$^+$ cations and RCOO$^-$ anions in neutral aqueous solution. For example, at room temperature, in a 1-molar solution of acetic acid, only 0.4% of the acid molecules are dissociated. Electronegative substituents give stronger acids.

Carboxylic acid	pK$_a$
Formic acid (HCOOH)	3.75
Acetic acid (CH$_3$COOH)	4.76
Chloroacetic acid (CH$_2$ClCO$_2$H)	2.86
Dichloroacetic acid (CHCl$_2$CO$_2$H)	1.29
Trichloroacetic acid (CCl$_3$CO$_2$H)	0.65
Trifluoroacetic acid (CF$_3$CO$_2$H)	0.23
Oxalic acid (HO$_2$CCO$_2$H)	1.27
Benzoic acid (C$_6$H$_5$CO$_2$H)	4.2

Deprotonation of carboxylic acids gives carboxylate anions; these are resonance stabilized, because the negative charge is delocalized over the two oxygen atoms, increasing the stability of the anion. Each of the carbon–oxygen bonds in the carboxylate anion has a partial double-bond character.

Odour

Carboxylic acids often have strong odors, especially the volatile derivatives. Most common are acetic acid (vinegar) and butyric acid (human vomit). Conversely esters of carboxylic acids tend to have pleasant odors and many are used in perfume.

Characterization

Carboxylic acids are readily identified as such by infrared spectroscopy. They exhibit a sharp band associated with vibration of the C–O vibration bond ($v_{C=O}$) between 1680 and 1725 cm^{-1}. A characteristic v_{O-H} band appears as a broad peak in the 2500 to 3000 cm^{-1} region. By ^{1}H NMR spectrometry, the hydroxyl hydrogen appears in the 10–13 ppm region, although it is often either broadened or not observed owing to exchange with traces of water.

Occurrence and Applications

Many carboxylic acids are produced industrially on a large scale. They are also pervasive in nature. Esters of fatty acids are the main components of lipids and polyamides of aminocarboxylic acids are the main components of proteins.

Carboxylic acids are used in the production of polymers, pharmaceuticals, solvents, and food additives. Industrially important carboxylic acids include acetic acid (component of vinegar, precursor to solvents and coatings), acrylic and methacrylic acids (precursors to polymers, adhesives), adipic acid (polymers), citric acid (beverages), ethylenediaminetetraacetic acid (chelating agent), fatty acids (coatings), maleic acid (polymers), propionic acid (food preservative), terephthalic acid (polymers).

Synthesis

Industrial Routes

In general, industrial routes to carboxylic acids differ from those used on smaller scale because they require specialized equipment.

- Oxidation of aldehydes with air using cobalt and manganese catalysts. The required aldehydes are readily obtained from alkenes by hydroformylation.

- Oxidation of hydrocarbons using air. For simple alkanes, the method is non-selective but too inexpensive to be useful. Allylic and benzylic compounds undergo more selective oxidations. Alkyl groups on a benzene ring oxidized to the carboxylic acid, regardless of its chain length. Benzoic acid from toluene, terephthalic acid from *para*-xylene, and phthalic acid from *ortho*-xylene are illustrative large-scale conversions. Acrylic acid is generated from propene.

- Base-catalyzed dehydrogenation of alcohols.

- Carbonylation is versatile method when coupled to the addition of water. This method is effective for alkenes that generate secondary and tertiary carbocations, e.g. isobutylene to pivalic acid. In the Koch reaction, the addition of water and carbon monoxide to alkenes is catalyzed by strong acids. Acetic acid and formic acid are produced by the carbonylation of methanol, conducted with iodide and alkoxide promoters, respectively, and often with high pressures of carbon monoxide, usually involving additional hydrolytic steps. Hydrocarboxylations involve the simultaneous addition of water and CO. Such reactions are sometimes called "Reppe chemistry":

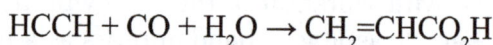

$$HCCH + CO + H_2O \rightarrow CH_2{=}CHCO_2H$$

- Some long-chain carboxylic acids are obtained by the hydrolysis of triglycerides obtained from plant or animal oils; these methods are related to soap making.

- fermentation of ethanol is used in the production of vinegar.

Laboratory Methods

Preparative methods for small scale reactions for research or for production of fine chemicals often employ expensive consumable reagents.

- oxidation of primary alcohols or aldehydes with strong oxidants such as potassium dichromate, Jones reagent, potassium permanganate, or sodium chlorite. The method is amenable to laboratory conditions compared to the industrial use of air, which is "greener", since it yields less inorganic side products such as chromium or manganese oxides.

- Oxidative cleavage of olefins by ozonolysis, potassium permanganate, or potassium dichromate.

- Carboxylic acids can also be obtained by the hydrolysis of nitriles, esters, or amides, in general with acid- or base-catalysis.

- Carbonation of a Grignard and organolithium reagents:

$$RLi + CO_2 \rightarrow RCO_2Li$$

$$RCO_2Li + HCl \rightarrow RCO_2H + LiCl$$

- Halogenation followed by hydrolysis of methyl ketones in the haloform reaction

- The Kolbe–Schmitt reaction provides a route to salicylic acid, precursor to aspirin.

Less-common Reactions

Many reactions afford carboxylic acids but are used only in specific cases or are mainly of academic interest:

- Disproportionation of an aldehyde in the Cannizzaro reaction

- Rearrangement of diketones in the benzilic acid rearrangement involving the generation of benzoic acids are the von Richter reaction from nitrobenzenes and the Kolbe–Schmitt reaction from phenols.

Reactions

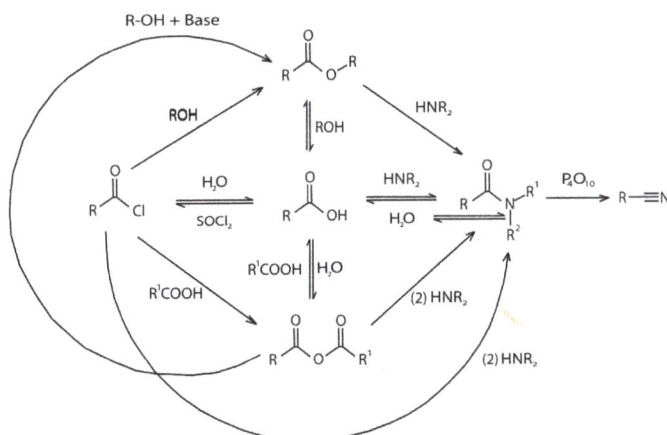

Carboxylic acid organic reactions

The most widely practiced reactions convert carboxylic acids into esters, amides, carboxylate salts, acid chlorides, and alcohols. Carboxylic acids react with bases to form carboxylate salts, in which the hydrogen of the hydroxyl (−OH) group is replaced with a metal cation. Thus, acetic acid found in vinegar reacts with sodium bicarbonate (baking soda) to form sodium acetate, carbon dioxide, and water:

$$CH_3COOH + NaHCO_3 \rightarrow CH_3COO^-Na^+ + CO_2 + H_2O$$

Carboxylic acids also react with alcohols to give esters. This process is heavily used in the production of polyesters. Likewise, carboxylic acids are converted into amides, but this conversion typically does not occur by direct reaction of the carboxylic acid and the amine. Instead esters are typical precursors to amides. The conversion of amino acids into peptides is a major biochemical process that requires ATP.

The hydroxyl group on carboxylic acids may be replaced with a chlorine atom using thionyl chloride to give acyl chlorides. In nature, carboxylic acids are converted to thioesters.

Carboxylic acid can be reduced to the alcohol by hydrogenation or using stoichiometric hydride reducing agents such as lithium aluminium hydride.

N,N-Dimethyl(chloromethylene)ammonium chloride (ClHC=N⁺(CH₃)₂Cl⁻) is a highly chemoselective agent for carboxylic acid reduction. It selectively activate the carboxylic acid and is known to tolerate active functionalities such as ketone as well as the moderate ester, olefin, nitrile, and halide moieties.

Specialized Reactions

- As with all carbonyl compounds, the protons on the α-carbon are labile due to keto–enol tautomerization. Thus, the α-carbon is easily halogenated in the Hell–Volhard–Zelinsky halogenation.

- The Schmidt reaction converts carboxylic acids to amines.

- Carboxylic acids are decarboxylated in the Hunsdiecker reaction.

- The Dakin–West reaction converts an amino acid to the corresponding amino ketone.

- In the Barbier–Wieland degradation, an carboxylic acid on an aliphatic chain having a simple the methylene bridge at the alpha position can have the chain shortened by one carbon. The inverse procedure is the Arndt–Eistert synthesis, where an acid is converted into acyl halide, which is then reacted with diazomethane to give one additional methylene in the aliphatic chain.

- Many acids undergo oxidative decarboxylation. Enzymes that catalyze these reactions are known as carboxylases and decarboxylases.

- Carboxylic acids are reduced to aldehydes via the ester and DIBAL, via the acid chloride in the Rosenmund reduction and via the thioester in the Fukuyama reduction.

- In ketonic decarboxylation carboxylic acids are converted to ketones.

- The Kolbe electrolysis is an electrolytic, decarboxylative dimerization reaction. In other words, it gets rid of the carboxyl groups of two acid molecules, and joins the remaining fragments together.

Amino Acid Dating

Amino acid dating is a dating technique used to estimate the age of a specimen in pale-obiology, molecular paleontology, archaeology, forensic science, taphonomy, sedimentary geology and other fields. This technique relates changes in amino acid molecules to the time elapsed since they were formed.

All biological tissues contain amino acids. All amino acids except glycine (the simplest one) are optically active, having a stereocenter at their α-C atom. This means that the amino acid can have two different configurations, "D" or "L" which are mirror images of each other. With a few important exceptions, living organisms keep all their amino acids in the "L" configuration. When an organism dies, control over the configuration of the amino acids ceases, and the ratio of D to L moves from a value near 0 towards an equilibrium value near 1, a process called racemization. Thus, measuring the ratio of D to L in a sample enables one to estimate how long ago the specimen died.

Factors Affecting Racemization

The rate at which racemization proceeds depends on the type of amino acid and on the average temperature, humidity, acidity (pH), and other characteristics of the enclosing matrix. Also, D/L concentration thresholds appear to occur as sudden decreases in the rate of racemization. These effects restrict amino acid chronologies to materials with known environmental histories and/or relative intercomparisons with other dating methods.

Temperature and humidity histories of microenvironments are being produced at ever increasing rates as technologies advance and technologists accumulate data. These are important for amino acid dating because racemization occurs much faster in warm, wet conditions compared to cold, dry conditions. Temperate to cold region studies are much more common than tropical studies, and the steady cold of the ocean floor or the dry interior of bones and shells have contributed most to the accumulation of racemization dating data. As a rule of thumb, sites with a mean annual temperature of 30°C have a maximum range of 200 ka and resolution of about 10 ka; sites at 10°C have a maximum age range of ~2 m.y., and resolution generally about 20% of the age; at -10°C the reaction has a maximum age of ~10 m.y., and a correspondingly coarser resolution.

Strong acidity and mild to strong alkalinity induce greatly increased racemization rates. Generally, they are not assumed to have a great impact in the natural environment, though tephrochronological data may shed new light on this variable.

The enclosing matrix is probably the most difficult variable in amino acid dating. This includes racemization rate variation among species and organs, and is affected by the depth of decomposition, porosity, and catalytic effects of local metals and minerals.

Amino Acids Used

Conventional racemization analysis tends to report a D-alloisoleucine / L-isoleucine (A/I or D/L ratio). This amino acid ratio has the advantages of being relatively easy to measure and being chronologically useful through the Quaternary.

Reverse phase HPLC techniques can measure up to 9 amino acids useful in geochronology over different time scales on a single chromatogram (aspartic acid, glutamic acid, serine, alanine, arginine, tyrosine, valine, phenylalanine, leucine).

In recent years there have been successful efforts to examine intra-crystalline amino acids separately as they have been shown to improve results in some cases.

Applications

Data from the geochronological analysis of amino acid racemization has been building for thirty-five years. Archeology, stratigraphy, oceanography, paleogeography, paleobiology, and paleoclimatology have been particularly affected. Their applications include dating correlation, relative dating, sedimentation rate analysis, sediment transport studies, conservation paleobiology, taphonomy and time-averaging, sea level determinations, and thermal history reconstructions.

Paleobiology and archaeology have also been strongly affected. Bone, shell, and sediment studies have contributed much to the paleontological record, including that relating to hominoids. Verification of radiocarbon and other dating techniques by amino acid racemization and vice versa has occurred. The 'filling in' of large probability ranges, such as with radiocarbon reservoir effects, has sometimes been possible. Paleopathology and dietary selection, paleozoogeography and indigineity, taxonomy and taphonomy, and DNA viability studies abound. The differentiation of cooked from uncooked bone, shell, and residue is sometimes possible. Human cultural changes and their effects on local ecologies have been assessed using this technique.

The slight reduction in this repair capability during aging is important to studies of longevity and old age tissue breakdown disorders, and allows the determination of age of living animals.

Amino acid racemization also has a role in tissue and protein degradation studies, particularly useful in developing museum preservation methods. These have produced models of protein adhesive and other biopolymer deteriorations and the concurrent pore system development.

Forensic science can use this technique to estimate the age of a cadaver or an objet d'art to determine authenticity.

Procedure

Amino acid racemization analysis consists of sample preparation, isolation of the amino acid wanted, and measure of its D:L ratio. Sample preparation entails the identification, raw extraction, and separation of proteins into their constituent amino acids, typically by grinding followed by acid hydrolysis. The amino acid derivative hydrolysis product can be combined with a chiral specific fluorescent, separated by chromatography or electrophoresis, and the particular amino acid D:L ratio determined by fluorescence. Alternatively, the particular amino acid can be separated by chromatography or electrophoresis, combined with a metal cation, and the D:L ratio determined by mass spectrometry. Chromatographic and electrophoretic separation of proteins and amino acids is dependent upon molecular size, which generally corresponds to molecular weight, and to a lesser extent upon shape and charge.

Nullomers

Nullomers are short sequences of DNA base pairs that do not occur in the genome of a species (commonly humans), even though they are theoretically possible. Nullomers must be under a selective pressure - for example, they may be toxic to the cell. Some nullomers have been shown to be useful to treat leukemia, breast, and prostate cancer. They are not useful in healthy cells because normal cells adapt and become immune to them. Nullomers are also being developed for use as DNA tags to prevent cross contamination when analyzing crime scene material.

Background

Nullomers are naturally available but potentially unused sequences of DNA. Determining these "forbidden" sequences can improve the understanding of the basic rules that govern sequence evolution. Sequencing the entire genome has shown that there is a high level of non-uniformity in genomic sequences. When a codon is artificially substituted for a synonymous codon, it often results in a lethal change and cell death. This is believed to be due to ribosomal stalling and early termination of protein synthesis. For example, both AGA and CGA code for arginine in bacteria; however, bacteria almost never use AGA, and when substituted it proves lethal. Such codon biases have been seen in all species, and are examples of constraints on sequence evolution. Other sequences may have selective pressure; for example, GG-rich sequences are used as sacrificial sinks for oxidative damage because oxidizing agents are attracted to regions with GG-rich sequences and then induce strand breakage.

Sequence of Human nullomers of 11bp in length	
No occurrence in the Human Genome	CGCTCGACGTA, GTCCGAGCGTA, CGACGAACGGT, CCGATACGTCG
One occurrence in the Human Genome	TACGCGCGACA, CGCGACGCATA, TCGGTACGCTA, TCGCGACCGTA, CGATCGTGCGA, CGCGTATCGGT
Two occurrences in the Human Genome	CGTCGCTCGAA, TCGCGCGAATA, TCGACGCGATA, ATCGTCGACGA, CTACGCGTCGA, CGTATACGCGA, CGATTACGCGA, CGATTCGGCGA, CGACGTACCGT, CGACGAACGAG, CGCGTAATACG, CGCGCTATACG
Three occurrences in the Human Genome	CGCGCATAATA, CGACGGCAGTA, CGAATCGCGTA, CGGTCGTACGA, GCGCGTACCGA, CGCGTAATCGA, CGTCGTTCGAC, CCGTCGAACGC, ACGCGCGATAT, CGAACGGTCGT, CGCGTAACGCG, CCGAATACGCG, CATATCGCGCG

Table of the number of nullomers present in different organisms and the nullomer length				
Organism	10bp	11bp	12bp	13bp
Arabidopsis	107	23646	1167012	20237388
C Elegans	2	7686	1152038	23339534
Chicken	2	590	131515	4722702
Chimpanzee	0	136	45938	2426474
Cow	0	96	45060	2432554
Dog	0	40	25217	1868964
Fruitfly	0	206	221616	12399300
Human	0	80	39852	2232448
Mouse	0	178	54383	2625646
Rat	0	50	30708	1933220
Zebrafish	0	2	15561	2469558

Cancer Treatment

Nullomers have been used as an approach to drug discovery and development. Nullomer peptides were screened for anti-cancer action. Absent sequences have short polyarginine tails added to increase solubility and uptake into the cell, producing peptides called PolyArgNulloPs. One successful sequence, RRRRRNWMWC, was demonstrated to have lethal effects in breast and prostate cancer. It damaged mitochondria by increasing ROS production, which reduced ATP production, leading to cell growth inhibition and cell death. Normal cells show a decreased sensitivity to PolyArgNulloPs over time.

Forensics

Accidental transfer of biological material containing DNA can produce misleading results. This is a particularly important consideration in forensic and crime labs, where mistakes can cause an innocent person to be convicted of a crime. There was no way to detect if a reference sample was mislabeled as evidence or if a forensic sample is contaminated, but a nullomer barcode can be added to reference samples to distinguish them from evidence on analysis. Tagging can be carried out during sample collection without affecting genotype or quantification results. Impregnated filter paper with various nullomers can be used to soak up and store DNA samples from a crime scene, making the technology simple and effective. Tagging with nullomers can be detected—even when diluted to a million-fold and spilled on evidence, these tags are still clearly detected. Tagging in this way supports National Research Council's recommendations on quality control to reduce fraud and mistakes.

References

- Jakubke, Hans-Dieter; Sewald, Norbert (2008). "Amino acids". Peptides from A to Z: A Concise Encyclopedia. Germany: Wiley-VCH. p. 20. ISBN 9783527621170 – via Google Books.

- Pollegioni, Loredano; Servi, Stefano, eds. (2012). Unnatural Amino Acids: Methods and Protocols. Methods in Molecular Biology – Volume 794. Humana Press. p. v. ISBN 978-1-61779-331-8. OCLC 756512314.

- Simon M (2005). Emergent computation: emphasizing bioinformatics. New York: AIP Press/ Springer Science+Business Media. pp. 105–106. ISBN 0-387-22046-1.

- Anfinsen CB, Edsall JT, Richards FM (1972). Advances in Protein Chemistry. New York: Academic Press. pp. 99, 103. ISBN 978-0-12-034226-6.

- Creighton, Thomas H. (1993). "Chapter 1". Proteins: structures and molecular properties. San Francisco: W. H. Freeman. ISBN 978-0-7167-7030-5.

- Liebecq, Claude, ed. (1992). Biochemical Nomenclature and Related Documents (2nd ed.). Portland Press. pp. 39–69. ISBN 978-1-85578-005-7.

- Smith, Anthony D. (1997). Oxford dictionary of biochemistry and molecular biology. Oxford: Oxford University Press. p. 535. ISBN 978-0-19-854768-6. OCLC 37616711.

- Simmons, William J.; Gerhard Meisenberg (2006). Principles of medical biochemistry. Mosby Elsevier. p. 19. ISBN 0-323-02942-6.

- Stryer, Lubert; Berg, Jeremy Mark; Tymoczko, John L. (2002). Biochemistry. San Francisco: W.H. Freeman. pp. 693–8. ISBN 0-7167-4684-0.

- Elmore, Donald Trevor; Barrett, G. C. (1998). Amino acids and peptides. Cambridge, UK: Cambridge University Press. pp. 48–60. ISBN 0-521-46827-2.

- McMurry, John (1996). Organic chemistry. Pacific Grove, CA, USA: Brooks/Cole. p. 1064. ISBN 0-534-23832-7.

- Jones, Russell Celyn; Buchanan, Bob B.; Gruissem, Wilhelm (2000). Biochemistry & molecular biology of plants. Rockville, Md: American Society of Plant Physiologists. pp. 371–2. ISBN 0-943088-39-9.

- Stryer, Lubert; Berg, Jeremy Mark; Tymoczko, John L. (2002). Biochemistry. San Francisco: W.H. Freeman. pp. 639–49. ISBN 0-7167-4684-0.

- Hausman, Robert E.; Cooper, Geoffrey M. (2004). The cell: a molecular approach. Washington, D.C: ASM Press. p. 51. ISBN 0-87893-214-3.

- Freifelder, D. (1983). Physical Biochemistry (2nd ed.). W. H. Freeman and Company. ISBN 0-7167-1315-2.

- McMurry, John E. (1992), Organic Chemistry (3rd ed.), Belmont: Wadsworth, ISBN 0-534-16218-5

- Lide, D. R., ed. (2005). CRC Handbook of Chemistry and Physics (86th ed.). Boca Raton (FL): CRC Press. ISBN 0-8493-0486-5.

- March, Jerry (1992), Advanced Organic Chemistry: Reactions, Mechanisms, and Structure (4th ed.), New York: Wiley, ISBN 0-471-60180-2

- Nelson, D. L.; Cox, M. M. (2000). Lehninger, Principles of Biochemistry (3rd ed.). New York: Worth Publishing. ISBN 1-57259-153-6.

- Haynes, William M., ed. (2011). CRC Handbook of Chemistry and Physics (92nd ed.). CRC Press. pp. 5–94 to 5–98. ISBN 1439855110.

Essential Amino Acid: An Overview

An essential amino acid is an amino acid that cannot be created by an organism from scratch. Histidine, isoleucine, leucine, methionine, tryptophan and valine are some of the topics explained in the text. The text strategically encompasses and incorporates the major components and key concepts of amino acids, providing a complete understanding.

Essential Amino Acid

An essential amino acid or indispensable amino acid is an amino acid that cannot be synthesized *de novo (from scratch)* by the organism, and thus must be supplied in its diet. The nine amino acids humans cannot synthesize are phenylalanine, valine, threonine, tryptophan, methionine, leucine, isoleucine, lysine, and histidine (i.e., F V T W M L I K H).

Six other amino acids are considered conditionally essential in the human diet, meaning their synthesis can be limited under special pathophysiological conditions, such as prematurity in the infant or individuals in severe catabolic distress. These six are arginine, cysteine, glycine, glutamine, proline and tyrosine (i.e. R C G Q P Y). Five amino acids are dispensable in humans, meaning they can be synthesized in the body. These five are alanine, aspartic acid, asparagine, glutamic acid and serine (i.e., A D N E S).

Essentiality in Humans

Essential	Nonessential
Histidine	Alanine
Isoleucine	Arginine*
Leucine	Aspartic acid
Lysine	Cysteine*
Methionine	Glutamic acid
Phenylalanine	Glutamine*
Threonine	Glycine*
Tryptophan	Proline*
Valine	Serine*
	Tyrosine*

	Asparagine*
	Selenocysteine

(*) Essential only in certain cases.

(**) Pyrrolysine, sometimes considered "the 22nd amino acid", is not listed here as it is not used by humans.

Eukaryotes can synthesize some of the amino acids from other substrates. Consequently, only a subset of the amino acids used in protein synthesis are essential nutrients.

Minimum Daily Intake

Estimating the daily requirement for the indispensable amino acids has proven to be difficult; these numbers have undergone considerable revision over the last 20 years. The following table lists the WHO recommended daily amounts currently in use for essential amino acids in adult humans, together with their standard one-letter abbreviations. Food sources are identified based on the USDA National Nutrient Database Release.

Amino acid(s)	mg per kg body weight	mg per 70 kg	mg per 100 kg
H Histidine	10	700	1000
I Isoleucine	20	1400	2000
L Leucine	39	2730	3900
K Lysine	30	2100	3000
M Methionine + C Cysteine	10.4 + 4.1 (15 total)	1050	1500
F Phenylalanine + Y Tyrosine	25 (total)	1750	2500
T Threonine	15	1050	1500
W Tryptophan	4	280	400
V Valine	26	1820	2600

The recommended daily intakes for children aged three years and older is 10% to 20% higher than adult levels and those for infants can be as much as 150% higher in the first year of life. Cysteine (or sulphur-containing amino acids), tyrosine (or aromatic amino acids), and arginine are always required by infants and growing children.

Relative Amino Acid Composition of Protein Sources

Various attempts have been made to express the "quality" or "value" of various kinds of protein. Measures include the biological value, net protein utilization, protein efficiency ratio, protein digestibility-corrected amino acid score and complete proteins concept. These concepts are important in the livestock industry, because the relative lack of one or more of the essential amino acids in animal feeds would have a limiting effect on growth and thus on feed conversion ratio. Thus, various feedstuffs may be fed in combination to increase net protein utilization, or a supplement of an individual amino acid (methionine, lysine, threonine, or tryptophan) can be added to the feed.

Although proteins from plant sources tend to have a relatively lower concentrations of protein by mass in comparison to protein from eggs or milk, they are nevertheless "complete" in that they contain at least trace amounts of all of the amino acids that are essential in human nutrition. Eating various plant foods in combination can provide a protein of higher biological value. Certain native combinations of foods, such as corn and beans, soybeans and rice, or red beans and rice, contain the essential amino acids necessary for humans in adequate amounts.

Additionally, certain types of algae and marine phytoplankton predate the division between animal and plant life on the planet; they have both chlorophyll as do plants, and also all the essential amino acids, as do animal proteins.

Protein Per Calorie

It can be shown that common vegetable sources contain adequate protein, often more protein per Calorie than the standard reference, whole raw egg, while other plant sources, particularly fruits contain less. For example, while 100 g of raw broccoli only provides 28 kcal and 3 g of protein, it has over 100 mg of protein per kcal. An egg contains five times as many calories (143 kcal) but only four times as much protein, roughly 90 mg of protein per kcal. However, a carrot has only 23 mg protein per kcal or twice the minimum recommendation, a banana meets the minimum, and an apple is below recommendation. It is recommended that adult humans obtain 10-35% of their calories as protein, or roughly 11–39 mg of protein per kcal per day (22-78 g for 2000 kcal). The US FDA daily reference value of 50 g protein per 2000 kcal is 25 mg/kcal per day.

Source	protein (g)	Calories (kcal)	protein/ Calorie (mg / kcal)	L (mg)	T (mg)	W (mg)	M+C (mg)
Apples, raw (100 g)	0.26	52	5	12	6	1	2
Minimum daily reference	22	2000	11				
Bananas, raw (100 g)	1	89	12	500	28	9	17
Carrot, raw (100 g)	1	41	23	101	191	12	103
US FDA daily / WHO (70 kg)	50	2000	25	2730	1050	280	1050
Upper daily reference	78	2000	39				
Peanut, valencia, raw (100 g)	48	570	84	1,627	859	244	630

Source	protein (g)	Calories (kcal)	protein/ Calorie (mg / kcal)	L (mg)	T (mg)	W (mg)	M+C (mg)
Soybeans, dry (100 g)	40	451	88	2634	1719	575	1172
Egg, whole, raw (100 g)	13	143	91	912	556	167	652
Broccoli, raw (100 g)	3	28	107	141	91	29	54
Soy Sauce, typical (100 g)	11	60	175	729	403	182	576
Beef, grass-fed, lean (100 g)	23	117	197				

Complete Proteins in Non-human Animals

Scientists had known since the early 20th century that rats could not survive on a diet whose only protein source was zein, which comes from maize (corn), but recovered if they were fed casein from cow's milk. This led William Cumming Rose to the discovery of the essential amino acid threonine. Through manipulation of rodent diets, Rose was able to show that ten amino acids are essential for rats: lysine, tryptophan, histidine, phenylalanine, leucine, isoleucine, methionine, valine, and arginine, in addition to threonine. Rose's later work showed that eight amino acids are essential for adult human beings, with histidine also being essential for infants. Longer term studies established histidine as also essential for adult humans.

Interchangeability

The distinction between essential and non-essential amino acids is somewhat unclear, as some amino acids can be produced from others. The sulfur-containing amino acids, methionine and homocysteine, can be converted into each other but neither can be synthesized *de novo* in humans. Likewise, cysteine can be made from homocysteine but cannot be synthesized on its own. So, for convenience, sulfur-containing amino acids are sometimes considered a single pool of nutritionally equivalent amino acids as are the aromatic amino acid pair, phenylalanine and tyrosine. Likewise arginine, ornithine, and citrulline, which are interconvertible by the urea cycle, are considered a single group.

Effects of Deficiency

If one of the nonessential amino acids is less than needed for an individual the utilization of other amino acids will be hindered and thus protein synthesis will be less than what it usually is, even in the presence of adequate total nitrogen intake.

Protein deficiency has been shown to affect all of the body's organs and many of its systems, including the brain and brain function of infants and young children; the immune system, thus elevating risk of infection; gut mucosal function and permeability, which affects absorption and vulnerability to systemic disease; and kidney function.

The physical signs of protein deficiency include edema, failure to thrive in infants and children, poor musculature, dull skin, and thin and fragile hair. Biochemical changes reflecting protein deficiency include low serum albumin and low serum transferrin.

The amino acids that are essential in the human diet were established in a series of experiments led by William Cumming Rose. The experiments involved elemental diets to healthy male graduate students. These diets consisted of cornstarch, sucrose, butterfat without protein, corn oil, inorganic salts, the known vitamins, a large brown "candy" made of liver extract flavored with peppermint oil (to supply any unknown vitamins), and mixtures of highly purified individual amino acids. The main outcome measure was nitrogen balance. Rose noted that the symptoms of nervousness, exhaustion, and dizziness were encountered to a greater or lesser extent whenever human subjects were deprived of an essential amino acid.

Essential amino acid deficiency should be distinguished from protein-energy malnutrition, which can manifest as marasmus or kwashiorkor. Kwashiorkor was once attributed to pure protein deficiency in individuals who were consuming enough calories ("sugar baby syndrome"). However, this theory has been challenged by the finding that there is no difference in the diets of children developing marasmus as opposed to kwashiorkor. Still, for instance in Dietary Reference Intakes (DRI) maintained by the USDA, lack of one or more of the essential amino acids is described as protein-energy malnutrition.

Histidine

Histidine (abbreviated as His or H; encoded by the codons CAU and CAC) is an α-amino acid that is used in the biosynthesis of proteins. It contains an α-amino group (which is in the protonated $-NH_3^+$ form under biological conditions), a carboxylic acid group (which is in the deprotonated $-COO^-$ form under biological conditions), and a side chain imidazole, classifying it as a positively charged amino acid at physiological pH. Initially thought essential only for infants, longer-term studies have shown it is essential for adults also.

Histidine was first isolated by German physician Albrecht Kossel and Sven Hedin in 1896. It is also a precursor to histamine, a vital inflammatory agent in immune responses. The acyl radical is histidyl.

Chemical Properties

The conjugate acid (protonated form) of the imidazole side chain in histidine has a pK_a of approximately 6.0. This means that, at physiologically relevant pH values, relatively small shifts in pH will change its average charge. Below a pH of 6, the imidazole ring is mostly protonated as described by the Henderson–Hasselbalch equation. When protonated, the imidazole ring bears two NH bonds and has a positive charge. The positive charge is equally distributed between both nitrogens and can be represented with two equally important resonance structures. As the pH increases past approximately 6, one of the protons is lost. The remaining proton of the now-neutral imidazole ring can reside on either nitrogen, giving rise to what are known as the N1-H or N3-H tautomers. The N3-H tautomer, shown in the figure above, is protonated on the #3 nitrogen, farther from the amino acid backbone bearing the amino and carboxyl groups, whereas the N1-H tautomer is protonated on the nitrogen nearer the backbone.

NMR and Tautomerism

When both imidazole ring nitrogens are protonated, their ^{15}N chemical shifts are similar (about 200 ppm, relative to nitric acid on the sigma scale, on which increased shielding corresponds to increased chemical shift). NMR shows that the chemical shift of N1-H drops slightly, whereas the chemical shift of N3-H drops considerably (about 190 vs. 145 ppm). This indicates that the N1-H tautomer is preferred, it is presumed due to hydrogen bonding to the neighboring ammonium. The shielding at N3 is substantially reduced due to the second-order paramagnetic effect, which involves a symmetry-allowed interaction between the nitrogen lone pair and the excited π^* states of the aromatic ring. As the pH rises above 9, the chemical shifts of N1 and N3 become approximately 185 and 170 ppm. An entirely deprotonated form of the imidazole ring, the imidazolate ion, would be formed only above a pH of 14, and is therefore not physiologically relevant. This change in chemical shifts can be explained by the presumably decreased hydrogen bonding of an amine over an ammonium ion, and the favorable hydrogen bonding between a carboxylate and an NH. This should act to decrease the N1-H tautomer preference.

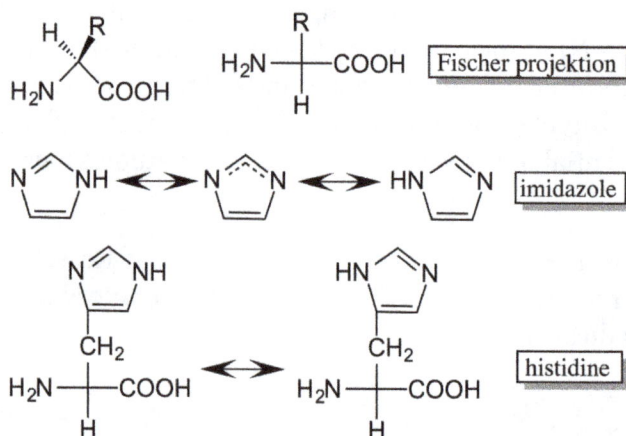

Various forms of histidine, showing the unphysiologic high-pH unprotonated form in the very center of the figure (it bears a negative charge, not shown, shared equally between the two nitrogens), and the neutral-pH singly protonated N3-H and N1-H tautomers at the lower left and right respectively. The physiologically relevant low-pH form, with two hydrogens and a positive charge shared equally between the two nitrogens, is not shown in this figure.

Aromaticity

The imidazole ring of histidine is aromatic at all pH values. It contains six pi electrons: four from two double bonds and two from a nitrogen lone pair. It can form pi stacking interactions, but is complicated by the positive charge. It does not absorb at 280 nm in either state, but does in the lower UV range more than some amino acids.

Biochemistry

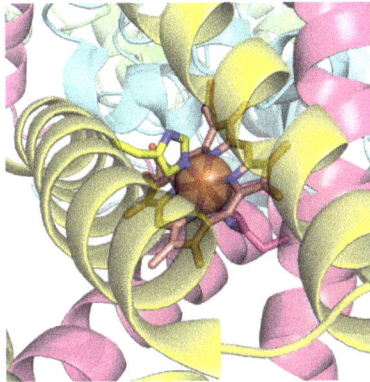

The histidine-bound heme group of succinate dehydrogenase, an electron carrier in the mitochondrial electron transfer chain. The large semi-transparent sphere indicates the location of the iron ion. From PDB: 1YQ3.

The imidazole sidechain of histidine is a common coordinating ligand in metalloproteins and is a part of catalytic sites in certain enzymes. It has the ability to switch between protonated and unprotonated states, which allows histidine to participate in acid-base catalysis. In catalytic triads, the basic nitrogen of histidine is used to abstract a proton from serine, threonine, or cysteine to activate it as a nucleophile. In a histidine proton shuttle, histidine is used to quickly shuttle protons. It can do this by abstracting a proton with its basic nitrogen to make a positively charged intermediate and then use another molecule, a buffer, to extract the proton from its acidic nitrogen. In carbonic anhydrases, a histidine proton shuttle is utilized to rapidly shuttle protons away from a zinc-bound water molecule to quickly regenerate the active form of the enzyme. Histidine is also important in haemoglobin in helices E and F. Histidine assists in stabilising oxyhaemoglobin and destabilising CO-bound haemoglobin. As a result, carbon monoxide binding is only 200 times stronger in haemoglobin, compared to 20,000 times stronger in free haem.

Metabolism

Biosynthesis

Histidine, also referred to as L-Histidine, is an essential amino acid that is not synthesized de novo in humans. Humans and other animals must ingest histidine or histidine-containing proteins. The biosynthesis of histidine has been widely studied in prokaryotes such as E. coli. Histidine synthesis in E.coli involves eight gene products (His1, 2, 3, 4, 5, 6, 7, and 8) and it occurs in ten steps. This is possible because a single gene product has the ability to catalyze more than one reaction. For example, as shown in the pathway, His4 catalyzes 4 different steps in the pathway.

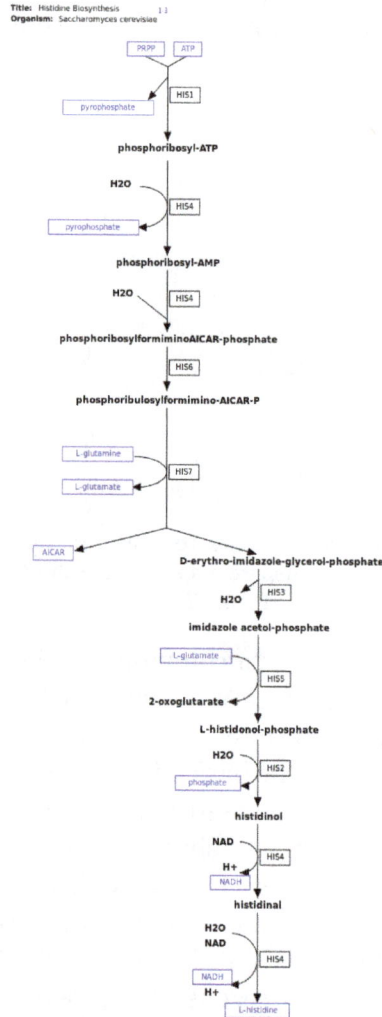

Histidine Biosynthesis Pathway Eight different enzymes can catalyze ten reactions. In this image, His4 catalyzes four different reactions in the pathway.

Histidine is synthesized from phosphoribosyl pyrophosphate (PRPP), a biochemical intermediate, which is made from ribose-5-phosphate by ribose-phosphate diphosphokinase during the pentose phosphate pathway. The first reaction of histidine biosynthesis is the condensation of PRPP and adenosine triphosphate (ATP) by the enzyme ATP-phosphoribosyl transferase. ATP-phosphoribosyl tranferase is indicated by His1 in the image. His4 gene product then hydrolyzes the product of the condensation, phosphoribosyl-ATP, producing phosphoribosyl-AMP (PRAMP), which is an irreversible step. His4 then catalyzes the formation of phosphoribosylformiminoAICAR-phosphate, which is then converted to phosphoribulosylformimino-AICAR-P by the His6 gene product. His7 splits phosphoribulosylformimino-AICAR-P to form D-erythro-imidazole-glycerol-phosphate. After, His3 forms imidazole acetol-phosphate releasing water. His5 then makes L-histidonol-phosphate, which is then hydrolyzed by His2 making histidonol. His4 catalyzes the oxidation of L-histidinol to form L-histidinal, an amino aldehyde. In the last step, L-histidinal is converted to L-histidine.

Just like animals and microorganisms, plants need histidine for their growth and development. Microorganisms and plants are similar in that they can synthesize histidine. Both synthesize histidine from the biochemical intermediate phosphoribosyl pyrophosphate. In general, the histidine biosynthesis is very similar in plants and microorganisms.

Regulation of Biosynthesis

This pathway requires energy in order to occur therefore, the presence of ATP activates the first enzyme of the pathway, ATP-phosphoribosyl transferase. ATP-phosphoribosyl transferase is the rate determining enzyme, which is regulated through feedback inhibition meaning that it is inhibited in the pres-ence of the product, histidine.

Degradation

Histidine is one of the amino acids that can be converted to intermediates of the tricarboxylic acid (TCA) cycle. Histidine along with other amino acids such as, proline and arginine, takes part in deamination, a process in which its amino group is removed. In prokaryptes, histidine is first converted to urocanate by histidase. Then, uracanase converts uracanate to 4-imidazolone-5-propionate. Imidazolonepropionase catalyzes the reaction to form formiminoglutamate (FIGLU) from 4-imidazolone-5-propionate. The formimino group is transferred to tetrahydrofolate, and the remaining five carbons form glutamate. Overall, these reactions result in the formation of glutamate and ammonia. Glutamate can then be deaminated by glutamate dehydrogenase or transaminated to form α-ketoglutarate.

Conversion to other Biologically Active Amines

- The histidine amino acid is a precursor for histamine, an amine produced in the body necessary for inflammation.

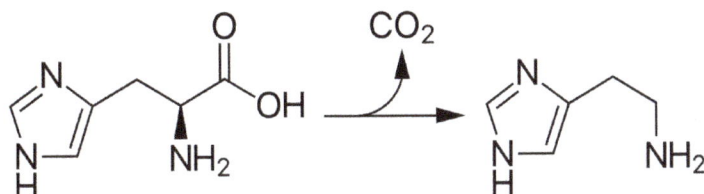

Conversion of histidine to histamine by histidine decarboxylase

- The enzyme histidine ammonia-lyase converts histidine into ammonia and urocanic acid. A deficiency in this enzyme is present in the rare metabolic disorder histidinemia, producing urocanic aciduria as a key diagnostic symptom.

- Histidine is also a precursor for carnosine biosynthesis, which is a dipeptide found in skeletal muscle.

- In Actinobacteria and filamentous fungi, such as *Neurospora crassa*, histidine can be converted into the antioxidant ergothioneine.

Supplementation

Supplementation of histidine has been shown to cause rapid zinc excretion in rats with an excretion rate 3 to 6 times normal.

Isoleucine

Isoleucine (abbreviated as Ile or I) encoded by the codons ATT, ATC, ATA is an α-amino acid that is used in the biosynthesis of proteins. It contains an α-amino group (which is in the protonated $-NH+3$ form under biological conditions), an α-carboxylic acid group (which is in the deprotonated $-COO^-$ form under biological conditions), and a hydrocarbon side chain, classifying it as a non-polar, uncharged(at physiological pH), aliphatic amino acid. It is essential in humans, meaning the body cannot synthesize it, and must be ingested in our diet. Isoleucine is synthesized from pyruvate employing leucine biosynthesis enzymes in other organisms such as bacteria.

Inability to break down isoleucine, along with other amino acids, is associated with the disease called Maple Syrup Urine Disease, which results in discoloration and a sweet smell in the patient's urine, which is where the name comes from. However in severe cases, MSUD can lead to damage to the brain cells and ultimately death.

Metabolism

Biosynthesis

As an essential nutrient, it is not synthesized in the body, hence it must be ingested, usually as a component of proteins. In plants and microorganisms, it is synthesized via several steps, starting from pyruvic acid and alpha-ketoglutarate. Enzymes involved in this biosynthesis include:

- Acetolactate synthase (also known as acetohydroxy acid synthase)

- Acetohydroxy acid isomeroreductase

- Dihydroxyacid dehydratase

- Valine aminotransferase

Catabolism

Isoleucine is both a glucogenic and a ketogenic amino acid. After transamination with alpha-ketoglutarate the carbon skeleton can be converted into either Succinyl CoA, and fed into the TCA cycle for oxidation or converted into oxaloacetate for gluconeogenesis (hence glucogenic). It can also be converted into Acetyl CoA and fed into the TCA cycle by condensing with oxaloacetate to form citrate. In mammals Acetyl CoA cannot be converted back to carbohydrate but can be used in the synthesis of ketone bodies or fatty acids, hence ketogenic.

Biotin, sometimes referred to as Vitamin B7 or Vitamin H, is an absolute requirement for the full catabolism of isoleucine (as well as leucine). Without adequate biotin, the human body will be unable to fully break down isoleucine and leucine molecules.

Nutritional Sources

Even though this amino acid is not produced in animals, it is stored in high quantities. Foods that have high amounts of isoleucine include eggs, soy protein, seaweed, turkey, chicken, lamb, cheese, and fish.

Isomers of Isoleucine

Forms of Isoleucine							
Common name:	isoleucine	D-isoleucine	L-isoleucine	DL-isoleucine	allo-D-isoleucine	allo-L-isoleucine	allo-DL-isoleucine
Synonyms:		(R)-Isoleucine	L(+)-Isoleucine	(R*,R*)-isoleucine		alloisoleucine	
PubChem:	CID 791 from PubChem	CID 94206 from PubChem	CID 6306 from PubChem	CID 76551 from PubChem			
EINECS number:	207-139-8	206-269-2	200-798-2		216-143-9	216-142-3	221-464-2
CAS number:	443-79-8	319-78-8	73-32-5		1509-35-9	1509-34-8	3107-04-8

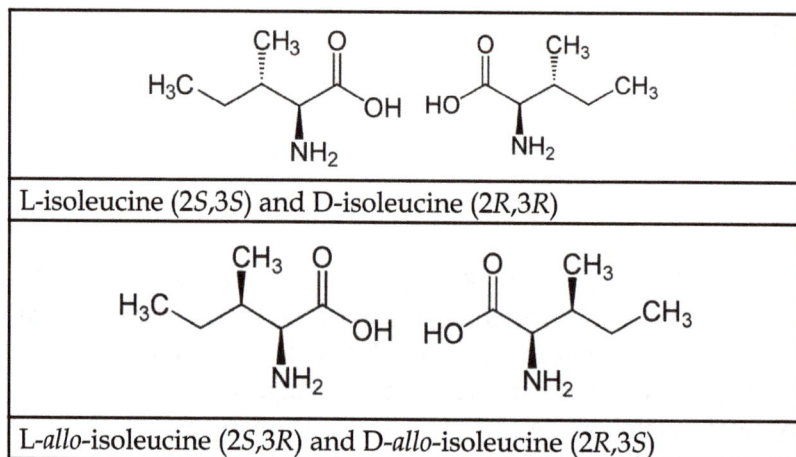

L-isoleucine (2*S*,3*S*) and D-isoleucine (2*R*,3*R*)

L-*allo*-isoleucine (2*S*,3*R*) and D-*allo*-isoleucine (2*R*,3*S*)

Synthesis

Isoleucine can be synthesized in a multistep procedure starting from 2-bromobutane and diethylmalonate. Synthetic isoleucine was originally reported in 1905.

German chemist Felix Ehrlich discovered isoleucine in hemoglobin in 1903.

Leucine

Leucine (abbreviated as Leu or L; encoded by the six codons UUA, UUG, CUU, CUC, CUA, and CUG) is an α-amino acid used in the biosynthesis of proteins. It contains an α-amino group (which is in the protonated $-NH+3$ form under biological conditions), an α-carboxylic acid group (which is in the deprotonated $-COO^-$ form under biological conditions), and an isobutyl side chain, classifying it as a nonpolar (at physiological pH) amino acid. It is essential in humans—meaning the body cannot synthesize it and thus must obtain from the diet.

Leucine is a major component of the subunits in ferritin, astacin, and other "buffer" proteins.

Biology

Leucine is used in the liver, adipose tissue, and muscle tissue. Adipose and muscle tissue use leucine in the formation of sterols. Combined leucine use in these two tissues is seven times greater than in the liver.

Biosynthesis in Plants

As it is an essential amino acid, animals cannot synthesize leucine. Consequently, they must ingest it, usually as a component of proteins. Plants and microorganisms synthesize leucine from pyruvic acid with a series of enzymes:

- Acetolactate synthase
- Acetohydroxy acid isomeroreductase
- Dihydroxyacid dehydratase
- α-Isopropylmalate synthase
- α-Isopropylmalate isomerase
- Leucine aminotransferase

Synthesis of the small, hydrophobic amino acid valine also includes the initial part of this pathway.

Metabolism in Humans

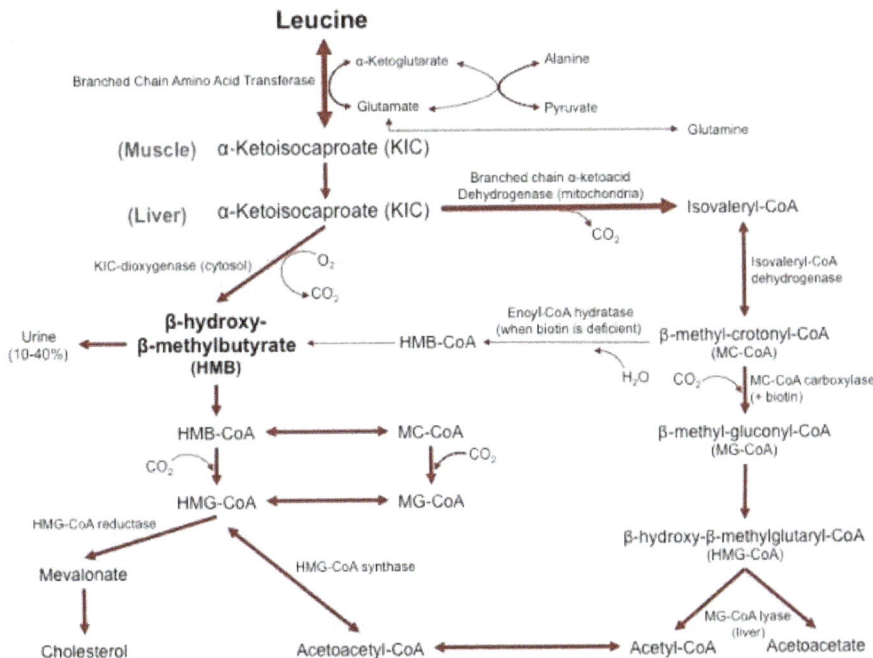

Human metabolic pathway for β-hydroxy β-methylbutyric acid (HMB) and isovaleryl-CoA, relative to L-leucine. Of the two major pathways, leucine is mostly metabolized into isovaleryl-CoA, while only about 5% is metabolized into HMB.

In healthy individuals, approximately 60% of dietary L-leucine is metabolized after several hours, with roughly 5% (2–10% range) of dietary L-leucine being converted to β-hydroxy β-methylbutyric acid (HMB). Around 40% of dietary L-leucine is converted to acetyl-CoA, which is subsequently used in the synthesis of other compounds.

The vast majority of L-leucine metabolism is initially catalyzed by the branched-chain amino acid aminotransferase enzyme, producing α-ketoisocaproate (α-KIC). α-Ketoisocaproate is mostly metabolized by the mitochondrial enzyme branched-chain α-ketoacid dehydrogenase, which converts it to isovaleryl-CoA. Isovaleryl-CoA is subsequently metabolized by isovaleryl-CoA dehydrogenase and converted to β-methylcrotonoyl-CoA (MC-CoA), which is used in the synthesis of acetyl-CoA and other compounds. During biotin deficiency, HMB can be synthesized from MC-CoA via enoyl-CoA hydratase and an unknown thioesterase enzyme, which convert MC-CoA into HMB-CoA and HMB-CoA into HMB respectively. A relatively small amount of α-KIC is metabolized in the liver by the cytosolic enzyme 4-hydroxyphenylpyruvate dioxygenase (*KIC dioxygenase*), which converts α-KIC to HMB. In healthy individuals, this minor pathway – which involves the conversion of L-leucine to α-KIC and then HMB – is the predominant route of HMB synthesis.

A small fraction of L-leucine metabolism – less than 5% in all tissues except the testes where it accounts for about 33% – is initially catalyzed by leucine aminomutase, producing β-leucine, which is subsequently metabolized into β-ketoisocaproate (β-KIC), β-ketoisocaproyl-CoA, and then acetyl-CoA by a series of uncharacterized enzymes. HMB could be produced via certain metabolites that are generated along this pathway, but as of 2015[j] the associated enzymes and reactions involved are not known.

The metabolism of HMB is initially catalyzed by an uncharacterized enzyme which converts it to HMB-CoA. HMB-CoA is metabolized by either enoyl-CoA hydratase or another uncharacterized enzyme, producing MC-CoA or hydroxymethylglutaryl-CoA (HMG-CoA) respectively. MC-CoA is then converted by the enzyme methylcrotonyl-CoA carboxylase to methylglutaconyl-CoA (MG-CoA), which is subsequently converted to HMG-CoA by methylglutaconyl-CoA hydratase. HMG-CoA is then cleaved into to acetyl-CoA and acetoacetate by HMG-CoA lyase or used in the production of cholesterol via the mevalonate pathway.

Effects

Leucine is an mTOR activator. It is a dietary amino acid with the capacity to directly stimulate muscle protein synthesis. As a dietary supplement, leucine has been found to slow the degradation of muscle tissue by increasing the synthesis of muscle proteins in aged rats. However, results of comparative studies are conflicted. Long-term leucine supplementation does not increase muscle mass or strength in healthy elderly men. More studies are needed, preferably ones based on an objective, random sample of society. Factors such as lifestyle choices, age, gender, diet, exercise, etc. must be factored into the analyses to isolate the effects of supplemental leucine as a standalone, or if taken with other branched chain amino acids (BCAAs). Until then, dietary supplemental leucine cannot be associated as the prime reason for muscular growth or optimal maintenance for the entire population.

Leucine potently activates the mammalian target of rapamycin kinase that regulates cell growth. Infusion of leucine into the rat brain has been shown to decrease food intake and body weight via activation of the mTOR pathway. Several sensing mechanisms have been proposed; most recently, it has been demonstrated that sestrin 2 can directly bind to leucine and activate mTORC1 activity by promoting its localization to the lysosome.

Both L-leucine and D-leucine protect mice against seizures. D-leucine also terminates seizures in mice after the onset of seizure activity, at least as effectively as diazepam and without sedative effects. Decreased dietary intake of L-leucine promotes adiposity in mice. High blood levels of leucine are associated with insulin resistance in humans, mice, and rodents.

Safety

Leucine toxicity, as seen in decompensated maple syrup urine disease (MSUD), causes delirium and neurologic compromise, and can be life-threatening.

Excess leucine may be a cause of pellagra, whose main symptoms are "the four D's": diarrhea, dermatitis, dementia and death, though the relationship is unclear.

Leucine at a dose exceeding 500 mg/kg/d was observed with hyperammonemia. As such, the UL for leucine in healthy adult men can be suggested at 500 mg/kg/d or 35 g/d under acute dietary conditions.

Dietary Sources

Food sources of leucine

Food	g/100g
Soybeans, mature seeds, roasted, salted	2.868
Hemp seed, hulled	2.163
Beef, round, top round, separable lean and fat, trimmed to 3 mm (⅛ in) fat, select, raw	1.76
Peanuts	1.672
Salami, pork	1.63
Fish, salmon, pink, raw	1.62
Wheat germ	1.571
Almonds	1.488
Chicken, broilers or fryers, thigh, meat only, raw	1.48
Chicken egg, yolk, raw, fresh	1.40
Oat	1.284
soybeans, Edamame, green, raw	0.926
Beans, pinto, cooked	0.765
Lentils, cooked	0.654
Chickpea, cooked	0.631
Corn, yellow	0.348

Food sources of leucine

Food	g/100g
Cow milk, whole, 3.25% milk fat	0.27
Rice, brown, medium-grain, cooked	0.191
Milk, human, mature, fluid	0.10

Chemical Properties

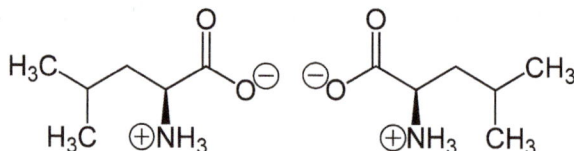

(S)-Leucine (or L-leucine), left; (R)-leucine (or D-leucine), right, in zwitterionic form at neutral pH

Leucine is a branched-chain amino acid (BCAA) since it possesses an aliphatic side-chain that is non-linear.

Racemic leucine had been subjected to circularly polarized synchrotron radiation to better understand the origin of biomolecular asymmetry. An enantiomeric enhancement of 2.6% had been induced, indicating a possible photochemical origin of biomolecules' homochirality.

Other Uses

As a food additive, L-leucine has E number E641 and is classified as a flavor enhancer.

Lysine

Lysine (abbreviated as Lys or K), encoded by the codons AAA and AAG) is an α-amino acid that is used in the biosynthesis of proteins. It contains an α-amino group (which is in the protonated $-NH_3^+$ form under biological conditions), an α-carboxylic acid group (which is in the deprotonated $-COO^-$ form under biological conditions), and a side chain lysyl ($(CH_2)_4NH_2$), classifying it as a charged (at physiological pH), aliphatic amino acid. It is essential in humans, meaning the body cannot synthesize it and thus it must be obtained from the diet.

Lysine is a base, as are arginine and histidine. The ε-amino group often participates in hydrogen bonding and as a general base in catalysis. The ε-amino group (NH_3^+) is attached to the fifth carbon from the α-carbon, which is attached to the carboxyl (C=OOH) group.

Common posttranslational modifications include methylation of the ε-amino group, giving methyl-, dimethyl-, and trimethyllysine (the latter occurring in calmodulin); also acetylation, sumoylation, ubiquitination, and hydroxylation - producing the hydroxylysine in collagen and other proteins. *O*-Glycosylation of hydroxylysine residues in the endoplasmic reticulum or Golgi apparatus is used to mark certain proteins for secretion from the cell. In opsins like rhodopsin and the visual opsins (encoded by the genes OPN1SW, OPN1MW, and OPN1LW), retinaldehyde forms a Schiff base with a conserved lysine residue, and interaction of light with the retinylidene group causes signal transduction in color vision. Deficiencies may cause blindness, as well as many other problems due to its ubiquitous presence in proteins.

Biosynthesis

As an essential amino acid, lysine is not synthesized in animals, hence it must be ingested as lysine or lysine-containing proteins. In plants and most bacteria, it is synthesized from aspartic acid (aspartate):

- L-aspartate is first converted to L-aspartyl-4-phosphate by aspartokinase (or Aspartate kinase). ATP is needed as an energy source for this step.

- β-Aspartate semialdehyde dehydrogenase converts this into β-aspartyl-4-semialdehyde (or β-aspartate-4-semialdehyde). Energy from NADPH is used in this conversion.

- 4-hydroxy-tetrahydrodipicolinate synthase adds a pyruvate group to the β-aspartyl-4-semialdehyde, and a water molecule is removed. This causes cyclization and gives rise to (2S,4S)-4-hydroxy-2,3,4,5-tetrahydrodipicolinate.

- This product is reduced to 2,3,4,5-tetrahydrodipicolinate (or Δ¹-piperidine-2,6-dicarboxylate, in the figure: (S)-2,3,4,5-tetrahydropyridine-2,6-dicarboxylate) by 4-hydroxy-tetrahydrodipicolinate reductase. This reaction consumes an NADPH molecule and releases a second water molecule.

- Tetrahydrodipicolinate N-acetyltransferase opens this ring and gives rise to N-succinyl-L-2-amino-6-oxoheptanedionate (or N-acyl-2-amino-6-oxopimelate). Two water molecules and one acyl-CoA (succinyl-CoA) enzyme are used in this reaction.

- N-succinyl-L-2-amino-6-oxoheptanedionate is converted into N-succinyl-LL-2,6-diaminoheptanedionate (N-acyl-2,6-diaminopimelate). This reaction is catalyzed by the enzyme succinyl diaminopimelate aminotransferase. A glutamic acid molecule is used in this reaction and an oxoacid is produced as a byproduct.

- N-succinyl-LL-2,6-diaminoheptanedionate (N-acyl-2,6-diaminopimelate) is converted into LL-2,6-diaminoheptanedionate (L,L-2,6-diaminopimelate) by succinyl diaminopimelate desuccinylase (acyldiaminopimelate deacylase). A water molecule is consumed in this reaction and a succinate is produced a byproduct.

- LL-2,6-diaminoheptanedionate is converted by diaminopimelate epimerase into meso-2,6-diamino-heptanedionate (meso-2,6-diaminopimelate).

- Finally, meso-2,6-diamino-heptanedionate is converted into L-lysine by diaminopimelate decarboxylase.

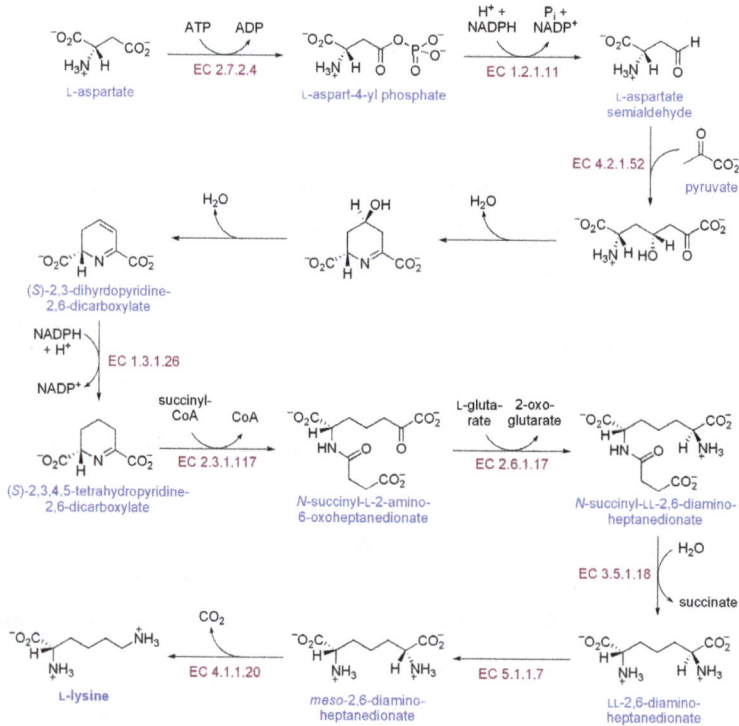

Enzymes involved in this biosynthesis include:

- Aspartokinase

- Aspartate-semialdehyde dehydrogenase

- 4-hydroxy-tetrahydrodipicolinate synthase

- 4-hydroxy-tetrahydrodipicolinate reductase

- 2,3,4,5-tetrahydropyridine-2,6-dicarboxylate N-succinyltransferase

- Succinyldiaminopimelate transaminase

- Succinyl-diaminopimelate desuccinylase

- Diaminopimelate epimerase

- Diaminopimelate decarboxylase.

It is worth noting, however, that in fungi, euglenoids and some prokaryotes lysine is synthesized via the alpha-aminoadipate pathway.

Metabolism

Lysine is metabolised in mammals to give acetyl-CoA, via an initial transamination with α-ketoglutarate. The bacterial degradation of lysine yields cadaverine by decarboxylation.

Allysine is a derivative of lysine, used in the production of elastin and collagen. It is produced by the actions of the enzyme lysyl oxidase on lysine in the extracellular matrix and is essential in the crosslink formation that stabilizes collagen and elastin.

Synthesis

Synthetic, racemic lysine has long been known. A practical synthesis starts from caprolactam. Industrially, L-lysine is usually manufactured by a fermentation process using *Corynebacterium glutamicum*; production exceeds 600,000 tons a year.

L-lysine HCl is used as a dietary supplement, providing 80.03% L-lysine. As such, 1 g of L-lysine is contained in 1.25 g of L-lysine HCl.

Dietary Sources

The nutritional requirement per day, in milligrams of lysine per kilogram of body weight, is: infants (3–4 months) 103 mg/kg, children (2 years) 64 mg/kg, older children (10–12 years) 44 to 60 mg/kg, adults 12 mg/kg. For a 70 kg adult, 12 milligrams of lysine per kilogram of body weight is 0.84 grams of lysine. Recommendations for adults have been revised upwards to 30 mg/kg.

Good sources of lysine are high-protein foods such as eggs, meat (specifically red meat, lamb, pork, and poultry), soy, beans and peas, cheese (particularly Parmesan), and certain fish (such as cod and sardines).

Lysine is the limiting amino acid (the essential amino acid found in the smallest quantity in the particular foodstuff) in most cereal grains, but is plentiful in most pulses (legumes). Consequently, meals that combine cereal grains and legumes, such as the Indian dal with rice, Middle Eastern hummus, ful medames, falafel with pita bread, the Mexican beans with rice or tortilla have arisen to provide complete protein in diets that are, by choice or by necessity, vegetarian. A food is considered to have sufficient lysine if it has at least 51 mg of lysine per gram of protein (so that the protein is 5.1% lysine).

Foods containing significant proportions of lysine include:

Food	Lysine (% of protein)	Notes
Catfish, channel, farmed, raw	9.19%	Bluefish, burbot, mahi-mahi, grouper, lingcod, mackerel, pike, salmon, scup, trout, tuna, and yellowtail also have lysine content of nearly 9.2%

Food	Lysine (% of protein)	Notes
Beef, ground, 90% lean/10% fat, cooked	8.31%	
Chicken, roasting, meat and skin, cooked, roasted	8.11%	
Lentil, sprouts, raw	7.95%	Sprouting increases the lysine content.
Parmesan cheese, grated	7.75%	
Azuki bean (adzuki beans), mature seeds, raw	7.53%	
Milk, non-fat	7.48%	
Soybean, mature seeds, raw	7.42%	
Pumpkin Seed, dried	7.4%	
Egg, whole, raw	7.27%	
Pea, split, mature seeds, raw	7.22%	
Winged bean (aka Goa Bean or Asparagus Pea), mature seeds, raw	7.20%	
Lentil, pink, raw	6.97%	
Kidney bean, mature seeds, raw	6.87%	
Chickpea, (garbanzo beans, Bengal gram), mature seeds, raw	6.69%	
Soybean, mature seeds, sprouts	5.74%	Sprouting decreases the lysine content.
Navy bean, mature seeds, raw	5.73%	
Amaranth, grain, uncooked	5.17%	
Quinoa	5%	

Properties

L-Lysine plays a major role in calcium absorption; building muscle protein; recovering from surgery or sports injuries; and the body's production of hormones, enzymes, and antibodies.

Modifications

Lysine can be modified through acetylation (acetyllysine), methylation (methyllysine), ubiquitination, sumoylation, neddylation, biotinylation, pupylation, and carboxylation, which tends to modify the function of the protein of which the modified lysine residue(s) are a part.

Clinical Significance

A systematic Cochrane Review (investigating all clinical trials, *in vitro* studies and mechanism of action) published in 2015 showed there is no evidence that lysine supplementation is effective against herpes simplex virus and it has not been approved by the FDA for herpes simplex suppression.

Lysine has anxiolytic action through its effects on serotonin receptors in the intestinal tract, and is also hypothesized to reduce anxiety through serotonin regulation in the amygdala. One study on rats showed that overstimulation of the 5-HT4 receptors in the gut are associated with anxiety-induced intestinal pathology. Lysine, acting as a serotonin antagonist and therefore reducing the overactivity of these receptors, reduced signs of anxiety and anxiety-induced diarrhea in the sample population. Another study showed that lysine deficiency leads to a pathological increase in serotonin in the amygdala, a brain structure that is involved in emotional regulation and the stress response. Human studies have also shown correlations between reduced lysine intake and anxiety. A population-based study in Syria included 93 families whose diet is primarily grain-based and therefore likely to be deficient in lysine. Fortification of grains with lysine was shown to reduce markers of anxiety, including cortisol levels; Smiriga and colleagues hypothesized that anxiety reduction from lysine occurs through mechanism of serotonin alterations in the central amygdala; older primary research reports hypothesized lysine to reduce anxiety through the potentiation of benzodiazepine receptors (common targets of anxiolytic drugs such as Xanax and Ativan).

There are lysine conjugates that show promise in the treatment of cancer, by causing cancerous cells to destroy themselves when the drug is combined with the use of phototherapy, while leaving non-cancerous cells unharmed.

Lysine deficiency causes immunodeficiency in chickens. One cause of relative lysine deficiency is cystinuria, where there is impaired hepatic resorption of basic, or positively charged amino acids, including lysine. The accompanying urinary cysteine results because the same deficient amino acid transporter is normally present in the kidney as well.

Limited studies suggest that a high-lysine diet or L-lysine monochloride supplements may have a moderating effect on blood pressure and the incidence of stroke.

Use of Lysine in Animal Feed

Lysine production for animal feed is a major global industry, reaching in 2009 almost 700,000 tonnes for a market value of over €1.22 billion. Lysine is an important additive to animal feed because it is a limiting amino acid when optimizing the growth of certain animals such as pigs and chickens for the production of meat. Lysine supplementation allows for the use of lower-cost plant protein (maize, for instance, rather than soy) while maintaining high growth rates, and limiting the pollution from nitrogen excretion. In turn, however, phosphate pollution is a major environmental cost when corn is used as feed for poultry and swine.

Lysine is industrially produced by microbial fermentation, from a base mainly of sugar. Genetic engineering research is actively pursuing bacterial strains to improve the efficiency of production and allow lysine to be made from other substrates.

In Popular Culture

The 1993 film *Jurassic Park* (based on the 1990 Michael Crichton novel of the same name) features dinosaurs that were genetically altered so that they could not produce lysine. This was known as the "lysine contingency" and was supposed to prevent the cloned dinosaurs from surviving outside the park, forcing them to be dependent on lysine supplements provided by the park's veterinary staff. In reality, no animals are capable of producing lysine (it is an essential amino acid).

In 1996, lysine became the focus of a price-fixing case, the largest in United States history. The Archer Daniels Midland Company paid a fine of US$100 million, and three of its executives were convicted and served prison time. Also found guilty in the price-fixing case were two Japanese firms (Ajinomoto, Kyowa Hakko) and a South Korean firm (Sewon). Secret video recordings of the conspirators fixing lysine's price can be found online or by requesting the video from the U.S. Department of Justice, Antitrust Division. This case served as the basis of the movie *The Informant!*, and a book of the same title.

Methionine

Methionine (abbreviated as Met or M; encoded by the codon AUG) is an α-amino acid that is used in the biosynthesis of proteins. It contains an α-amino group (which is in the protonated $-NH_3^+$ form under biological conditions), an α-carboxylic acid group (which is in the deprotonated $-COO^-$ form under biological conditions), and an S-methyl thioether side chain, classifying it as a non-polar, aliphatic amino acid. It is essential in humans, meaning the body cannot synthesize it and thus it must be obtained from the diet.

Methionine is coded for by the initiation codon meaning it indicates the start of the coding region and is the first amino acid produced in a nascent polypeptide during mRNA translation.

Methionine: A Proteinogenic Amino Acid

Together with cysteine, methionine is one of two sulfur-containing proteinogenic amino acids. Excluding the few exceptions where methionine may act as a redox sensor (*e.g.*), methionine residues do not have a catalytic role. This is in contrast to cysteine residues, where the thiol group has a catalytic role in many proteins. The thioether does however have a minor structural role due to the stability effect of S/π interactions between the side chain sulfur atom and aromatic amino acids in one-third of all known protein structures. This lack of a strong role is reflected in experiments where little effect is seen in proteins where methionine is replaced by norleucine, a straight hydrocarbon sidechain amino acid which lacks the thioether. It has been conjectured that norleucine was present in early versions of the genetic code, but methionine intruded into the final version of the genetic code due to the fact it is used in the cofactor *S*-adenosyl methionine (SAM). This situation is not unique and may have occurred with ornithine and arginine.

Encoding

Methionine is one of only two amino acids encoded by a single codon (AUG) in the standard genetic code (tryptophan, encoded by UGG, is the other). In reflection to the evolutionary origin of its codon, the other AUN codons encode isoleucine, which is also a hydrophobic amino acid. In the mitochondrial genome of several organisms, including metazoa and yeast, the codon AUA also encodes for methionine. In the standard genetic code AUA codes for isoleucine and the respective tRNA (*ileX* in *Escherichia coli*) uses the unusual base lysidine (bacteria) or agmatine (archaea) to discriminate against AUG.

The methionine codon AUG is also the most common start codon. A "Start" codon is message for a ribosome that signals the initiation of protein translation from mRNA when the AUG codon is in a Kozak consensus sequence. As a consequence, methionine is often incorporated into the N-terminal position of proteins in eukaryotes and archaea during translation, although it can be removed by post-translational modification. In bacteria, the derivative N-formylmethionine is used as the initial amino acid.

Methionine Derivatives

S-adenosyl-methionine

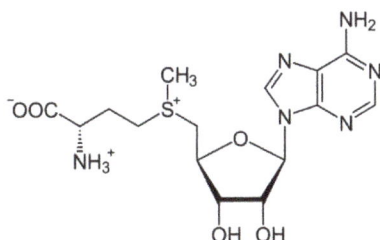

S-Adenosyl-methionine is a cofactor derived from methionine.

The methionine-derivative *S*-adenosyl methionine (SAM) is a cofactor that serves mainly as a methyl donor. SAM is composed of an adenosyl molecule (via 5' carbon) attached to the sulfur of methionine, therefore making it a sulfonium cation (*i.e.* three substituents and positive charge). The sulfur acts as soft Lewis acid (*i.e.* donor/electrophile) allows the *S*-methyl group to be transferred to an oxygen, nitrogen or aromatic system, often with the aid of other cofactors such as cobalamin (vitamin B12 in humans). Some enzymes use SAM to initiate a radical reaction, these are called radical SAM enzymes. As a result of the transfer of the methyl group, S-adenosyl-homocysteine is obtained. In bacteria, this is either regenerated by methylation or is salvaged by removing the adenine and the homocysteine leaving the compound dihydroxypentandione to spontaneously convert into autoinducer-2, which is excreted as a waste product / quorum signal.

Biosynthesis

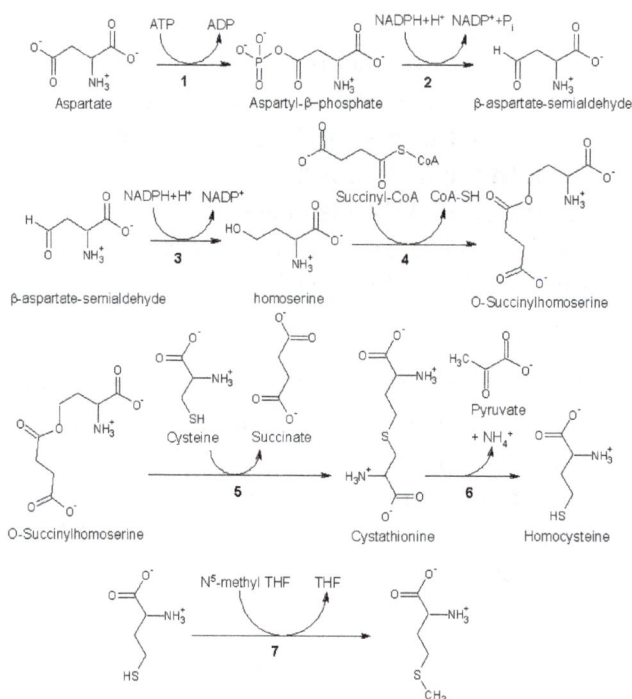

Methionine biosynthesis

As an essential amino acid, methionine is not synthesized de novo in humans and other animals, who must ingest methionine or methionine-containing proteins. In plants and microorganisms, methionine biosynthesis belongs to the aspartate family, along with threonine and lysine (via diaminopimelate, but not via α-aminoadipate). The main backbone is derived from aspartic acid, while the sulfur may come from cysteine, methanethiol or hydrogen sulfide.

- First, aspartic acid is converted via β-aspartyl-semialdehyde into homoserine by two reduction steps of the terminal carboxyl group (homoserine has therefore a γ-hydroxyl, hence the homo- series). The intermediate aspartate-semialdehyde is the branching point with the lysine biosynthetic pathway, where it is instead condensed with pyruvate. Homoserine is the branching point with the threonine pathway, where instead it is isomerised after activating the termainal hydroxyl with phosphate (also used for methionine biosynthesis in plants).

- Homoserine is then activated with an phosphate, succinyl or acetyl gorup on the hydroxyl.

  - In plants and possibly in some bacteria (viz.) phosphate is used. This step is shared with threonine biosynthesis.

  - In most organisms, an acetyl group is used to activate the homoserine. This can be catalysed in bacteria by an enzyme encoded by *metX* or *metA* (not homologues).

  - In enterobacteria and a limited amount of other organisms, succinate is used. The enzyme that catalyses the reaction is MetA and the specificity for acetyl-CoA and succinyl-CoA is dictated by a single reside. The physiological basis for the preference of acetyl-CoA or succinyl-CoA is unknown, but such alternative routes are present in some other pathways (*e.g.* lysine biosynthesis and arginine biosynthesis).

- The hydroxyl activating group is then replaced with cysteine, methanethiol or hydrogen sulfide. A replacement reaction is technically a γ-elimination followed by a variant of a Michael addition. All the enzymes involved are homologues and members of the Cys/Met metabolism PLP-dependent enzyme family, which is a subset of the PLP-dependant fold type I clade. They utilise the cofactor PLP (pyridoxal phosphate), which functions by stabilising carbanion intermediates.

  - If it reacts with cysteine, it produces cystathionine, which is cleaved to yield homocysteine. The enzymes involved are cystathionine-γ-synthase (encoded by *metB* in bacteria) and cystathionine-β-lyase (*metC*). Cystathionine is bound differently in the two enzymes allowing β or γ reactions to occur.

  - If it reacts with free hydrogen sulfide, it produces homocysteine. This is catalysed by *O*-acetylhomoserine aminocarboxypropyltransferase (formerly known as *O*-acetylhomoserine (thiol)-lyase. It is encoded by either *metY* or *metZ* in bacteria.

  - If it reacts with methanethiol, it produces methionine directly. Methanethiol is a byproduct of catabolic pathway of certain compounds, therefore this route is more uncommon.

- If homocysteine is produced the thiol group is methylated, yielding methionine. Two methionine synthases are known, one cobalamin (vitamin B_{12}) dependent and one independent.

The pathway utilising cysteine is called the "Transsulfuration pathway", while the pathway utilising hydrogen sulfide (or methanethiol) is called "direct-sulfurylation pathway".

Cysteine is similarly produced, namely it can be made from an activated serine and either from homocysteine ("reverse trans-sulfurylation route") or from hydrogen sulfide ("direct sulfurylation route"); the activated serine is generally O-acetyl-serine (via CysK or CysM in $E.$ $coli$), but in Aeropyrum pernix and some other archaea O-phosphoserine is used. CysK and CysM are homologues, but belong to the PLP fold type III clade.

Trans-sulfurylation Pathway

Enzymes involved in the $E.$ $coli$ trans-sulfurylation route of methionine biosynthesis:

1. Aspartokinase

2. Aspartate-semialdehyde dehydrogenase

3. Homoserine dehydrogenase

4. Homoserine O-transsuccinylase

5. Cystathionine-γ-synthase

6. Cystathionine-β-lyase

7. Methionine synthase (in mammals, this step is performed by Homocysteine methyltransferase or Betaine—homocysteine S-methyltransferase)

Other Biochemical Pathways

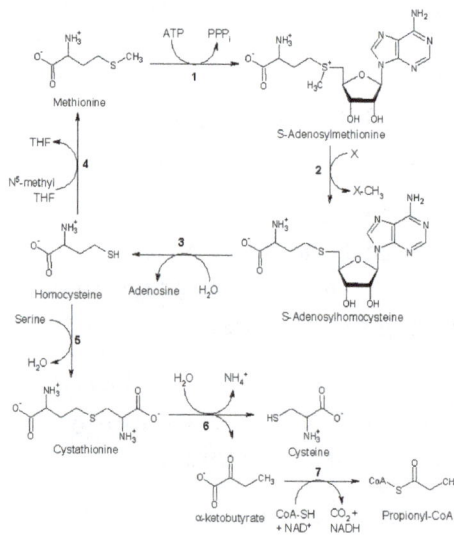

Fates of methionine

Although mammals cannot synthesize methionine, they can still use it in a variety of biochemical pathways:

Methionine Catabolism

Methionine is converted to S-adenosylmethionine (SAM) by (1) methionine adenosyl-transferase.

SAM serves as a methyl-donor in many (2) methyltransferase reactions, and is converted to *S*-adenosylhomocysteine (SAH).

(3) Adenosylhomocysteinase converts SAH to homocysteine.

There are two fates of homocysteine: it can be used to regenerate methionine, or to form cysteine.

Regeneration of Methionine

Methionine can be regenerated from homocysteine via (4) methionine synthase in a reaction that requires Vitamin B_{12} as a cofactor.

Homocysteine can also be remethylated using glycine betaine (NNN-trimethyl glycine, TMG) to methionine via the enzyme betaine-homocysteine methyltransferase (E.C.2.1.1.5, BHMT). BHMT makes up to 1.5% of all the soluble protein of the liver, and recent evidence suggests that it may have a greater influence on methionine and homocysteine homeostasis than methionine synthase.

Reverse-transulfurylation Pathway: Conversion to Cysteine

Homocysteine can be converted to cysteine.

- (5) Cystathionine-β-synthase (a PLP-dependent enzyme) combines homocysteine and serine to produce cystathionine. Instead of degrading cystathionine via cystathionine-β-lyase, as in the biosynthetic pathway, cystathionine is broken down to cysteine and α-ketobutyrate via (6) cystathionine-γ-lyase.

- (7) The enzyme α-ketoacid dehydrogenase converts α-ketobutyrate to propionyl-CoA, which is metabolized to succinyl-CoA in a three-step process.

Ethylene Synthesis

This amino acid is also used by plants for synthesis of ethylene. The process is known as the Yang Cycle or the methionine cycle.

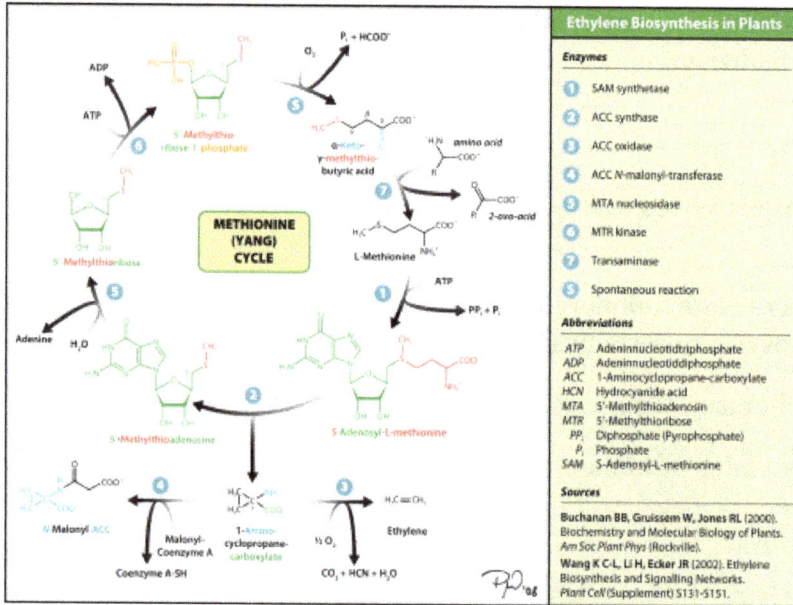

The Yang cycle

Chemical Synthesis

Racemic methionine can be synthesized from diethyl sodium phthalimidomalonate by alkylation with chloroethylmethylsulfide ($ClCH_2CH_2SCH_3$) followed by hydrolysis and decarboxylation.

Human Nutrition Dietary

Sources

Food sources of Methionine

Food	g/100g
Egg, white, dried, powder, glucose reduced	3.204
Sesame seeds flour (low fat)	1.656
Egg, whole, dried	1.477
Cheese, Parmesan, shredded	1.114
Brazil nuts	1.008
Soy protein concentrate	0.814
Chicken, broilers or fryers, roasted	0.801
Fish, tuna, light, canned in water, drained solids	0.755
Beef, cured, dried	0.749
Bacon	0.593
Beef, ground, 95% lean meat / 5% fat, raw	0.565
Pork, ground, 96% lean / 4% fat, raw	0.564

Food sources of Methionine

Food	g/100g
Wheat germ	0.456
Oat	0.312
Peanuts	0.309
Chickpea	0.253
Corn, yellow	0.197
Almonds	0.151
Beans, pinto, cooked	0.117
Lentils, cooked	0.077
Rice, brown, medium-grain, cooked	0.052

High levels of methionine can be found in eggs, sesame seeds, Brazil nuts, fish, meats and some other plant seeds; methionine is also found in cereal grains. Most fruits and vegetables contain very little of it. Most legumes are also low in methionine. However, it is the combination of methionine and cystine which is considered for completeness of a protein. Racemic methionine is sometimes added as an ingredient to pet foods.

Methionine Restriction

There is scientific evidence that restricting methionine consumption can increase lifespans in some animals.

A 2005 study showed methionine restriction without energy restriction extends mouse lifespan.

A study published in *Nature* showed adding just the essential amino acid methionine to the diet of fruit flies under dietary restriction, including restriction of essential amino acids (EAAs), restored fertility without reducing the longer lifespans that are typical of dietary restriction, leading the researchers to determine that methionine "acts in combination with one or more other EAAs to shorten lifespan."

Several studies showed that methionine restriction also inhibits aging-related disease processes in mice and inhibits colon carcinogenesis in rats. In humans, methionine restriction through dietary modification could be achieved through a vegan diet. Veganism being a completely plant based diet is typically very low in methionine, however certain nuts and legumes may provide higher levels.

A 2009 study on rats showed "methionine supplementation in the diet specifically increases mitochondrial ROS production and mitochondrial DNA oxidative damage in rat liver mitochondria offering a plausible mechanism for its hepatotoxicity".

However, since methionine is an essential amino acid, it cannot be entirely removed from animals' diets without disease or death occurring over time. For example, rats fed a diet without methionine developed steatohepatitis (fatty liver), anemia and lost two thirds of their body weight over 5 weeks. Administration of methionine ameliorated the pathological consequences of methionine deprivation.

Methionine might also be essential to reversing damaging methylation of glucocorticoid receptors caused by repeated stress exposures, with implications for depression.

Health

Loss of methionine has been linked to senile greying of hair. Its lack leads to a buildup of hydrogen peroxide in hair follicles, a reduction in tyrosinase effectiveness, and a gradual loss of hair color.

Methionine is an intermediate in the biosynthesis of cysteine, carnitine, taurine, lecithin, phosphatidylcholine, and other phospholipids. Improper conversion of methionine can lead to atherosclerosis.

Other Uses

DL-Methionine is sometimes given as a supplement to dogs; It helps to reduce the chances of stones in dogs. Methionine is also known to increase the urinary excretion of quinidine by acidifying the urine. Aminoglycoside antibiotics used to treat urinary tract infections work best in alkaline conditions, and urinary acidification from using methionine can reduce its effectiveness. If a dog is on a diet that acidifies the urine, methionine should not be used.

Methionine is allowed as a supplement to organic poultry feed under the US certified organic program.

Phenylalanine

Phenylalanine is an α-amino acid with the formula $C_9H_{11}NO_2$. It can be viewed as a benzyl group substituted for the methyl group of alanine, or a phenyl group in place of a terminal hydrogen of alanine. This essential amino acid is classified as neutral, and nonpolar because of the inert and hydrophobic nature of the benzyl side chain. The L-isomer is used to biochemically form proteins, coded for by DNA. The codons for L-phenylalanine are UUU and UUC. Phenylalanine is a precursor for tyrosine; the monoamine neurotransmitters dopamine, norepinephrine (noradrenaline), and epinephrine (adrenaline); and the skin pigment melanin.

Phenylalanine is found naturally in the breast milk of mammals. It is used in the manufacture of food and drink products and sold as a nutritional supplement for its reputed analgesic and antidepressant effects. It is a direct precursor to the neuromodulator phenethylamine, a commonly used dietary supplement.

History

The first description of phenylalanine was made in 1879, when Schulze and Barbieri identified a compound with the empirical formula, $C_9H_{11}NO_2$, in yellow lupine (*Lupinus luteus*) seedlings. In 1882, Erlenmeyer and Lipp first synthesized phenylalanine from phenylacetaldehyde, hydrogen cyanide, and ammonia.

The genetic codon for phenylalanine was first discovered by J. Heinrich Matthaei and Marshall W. Nirenberg in 1961. They showed that by using mRNA to insert multiple uracil repeats into the genome of the bacterium *E. coli*, they could cause the bacterium to produce a polypeptide consisting solely of repeated phenylalanine amino acids. This discovery helped to establish the nature of the coding relationship that links information stored in genomic nucleic acid with protein expression in the living cell.

Biosynthesis

As an essential amino acid, phenylalanine is not synthesized de novo in humans and other animals, who must ingest phenylalanine or phenylalanine-containing proteins.

Dietary Sources

Good sources of phenylalanine are eggs, chicken, liver, beef, milk, and soybeans. Other sources include spinach and leafy greens, tofu, amaranth leaves, and lupin seeds.

Other Biological Roles

L-Phenylalanine is biologically converted into L-tyrosine, another one of the DNA-encoded amino acids. L-tyrosine in turn is converted into L-DOPA, which is further converted into dopamine, norepinephrine (noradrenaline), and epinephrine (adrenaline). The latter three are known as the catecholamines.

Phenylalanine uses the same active transport channel as tryptophan to cross the blood–brain barrier. In excessive quantities, supplementation can interfere with the production of serotonin and other aromatic amino acids as well as nitric oxide due to the overuse (eventually, limited availability) of the associated cofactors, iron or tetrahydrobiopterin. The corresponding enzymes in for those compounds are the aromatic amino acid hydroxylase family and nitric oxide synthase.

Biosynthetic pathways for catecholamines and trace amines in the human brain

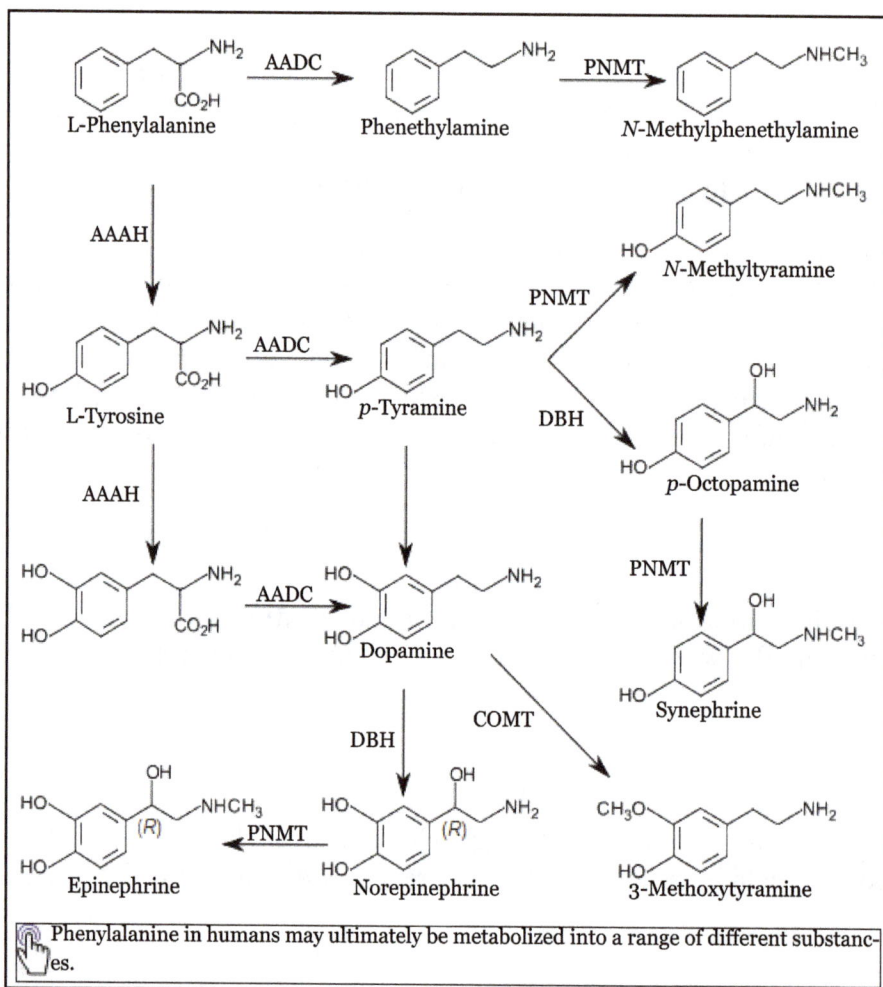

Phenylalanine in humans may ultimately be metabolized into a range of different substances.

In Plants

Phenylalanine is the starting compound used in the synthesis of flavonoids. Lignan is derived from phenylalanine and from tyrosine. Phenylalanine is converted to cinnamic acid by the enzyme phenylalanine ammonia-lyase.

Phenylketonuria

The genetic disorder phenylketonuria (PKU) is the inability to metabolize phenylalanine because of a lack of the enzyme phenylalanine hydroxylase. Individuals with this disorder are known as "phenylketonurics" and must regulate their intake of phenylalanine. A (rare) "variant form" of phenylketonuria called hyperphenylalaninemia is caused by the inability to synthesize a cofactor called tetrahydrobiopterin, which can be supplemented. Pregnant women with hyperphenylalaninemia may show similar symptoms of the disorder (high levels of phenylalanine in blood) but these indicators will usually disappear at the end of gestation. Pregnant women with PKU must control their blood phenylalanine levels even if the fetus is heterozygus for the defective gene because the fetus could be adversely affected due to hepatic immaturity. Individuals who cannot metabolize phenylalanine must monitor their intake of protein to control the buildup of phenylalanine as their bodies convert protein into its component amino acids.

Phenylketonurics often use blood tests to monitor the amount of phenylalanine in their blood. Lab results may report phenylalanine levels in different units, including mg/dL and μmol/L. One mg/dL of phenylalanine is approximately equivalent to 60 μmol/L.

A non-food source of phenylalanine is the artificial sweetener aspartame. This compound, sold under the trade names Equal and NutraSweet, is metabolized by the body into several chemical byproducts including phenylalanine. The breakdown problems phenylketonurics have with protein and the attendant buildup of phenylalanine in the body also occurs with the ingestion of aspartame, although to a lesser degree. Accordingly, all products in Australia, the U.S. and Canada that contain aspartame must be labeled: "Phenylketonurics: Contains phenylalanine." In the UK, foods containing aspartame must carry ingredient panels that refer to the presence of "aspartame or E951" and they must be labeled with a warning "Contains a source of phenylalanine." In Brazil, the label "Contém Fenilalanina" (Portuguese for "Contains Phenylalanine") is also mandatory in products which contain it. These warnings are placed to aid individuals who have been diagnosed with PKU so that they can avoid such foods.

Geneticists have recently sequenced the genome of macaques. Their investigations have found "some instances where the normal form of the macaque protein looks like the diseased human protein" including markers for PKU.

D-, L- and DL-phenylalanine

The stereoisomer D-phenylalanine (DPA) can be produced by conventional organic synthesis, either as a single enantiomer or as a component of the racemic mixture. It does not participate in protein biosynthesis although it is found in proteins in small amounts - particularly aged proteins and food proteins that have been processed. The biological functions of D-amino acids remain unclear, although D-phenylalanine has pharmacological activity at niacin receptor 2.

DL-Phenylalanine (DLPA) is marketed as a nutritional supplement for its supposed analgesic and antidepressant activities. DL-Phenylalanine is a mixture of D-phenylalanine and L-phenylalanine. The reputed analgesic activity of DL-phenylalanine may be explained by the possible blockage by D-phenylalanine of enkephalin degradation by the enzyme carboxypeptidase A. The mechanism of DL-phenylalanine's supposed antidepressant activity may be accounted for by the precursor role of L-phenylalanine in the synthesis of the neurotransmitters norepinephrine and dopamine. Elevated brain levels of norepinephrine and dopamine are thought to have an antidepressant effect. D-Phenylalanine is absorbed from the small intestine and transported to the liver via the portal circulation. A small amount of D-phenylalanine appears to be converted to L-phenylalanine. D-Phenylalanine is distributed to the various tissues of the body via the systemic circulation. It appears to cross the blood–brain barrier less efficiently than L-phenylalanine, and so a small amount of an ingested dose of D-phenylalanine is excreted in the urine without penetrating the central nervous system.

L-Phenylalanine is an antagonist at $\alpha 2\delta$ Ca^{2+} calcium channels with a K_i of 980 nM.

In the brain, L-phenylalanine is a competitive antagonist at the glycine binding site of NMDA receptor and at the glutamate binding site of AMPA receptor. At the glycine binding site of NMDA receptor L-phenylalanine has an apparent equilibrium dissociation constant (K_B) of 573 µM estimated by Schild regression which is considerably lower than brain L-phenylalanine concentration observed in untreated human phenylketonuria. L-Phenylalanine also inhibits neurotransmitter release at glutamatergic synapses in hippocampus and cortex with IC_{50} of 980 µM, a brain concentration seen in classical phenylketonuria, whereas D-phenylalanine has a significantly smaller effect.

Commercial Synthesis

L-Phenylalanine is produced for medical, feed, and nutritional applications, such as aspartame, in large quantities by utilizing the bacterium *Escherichia coli*, which naturally produces aromatic amino acids like phenylalanine. The quantity of L-phenylalanine produced commercially has been increased by genetically engineering *E. coli*, such as by altering the regulatory promoters or amplifying the number of genes controlling enzymes responsible for the synthesis of the amino acid.

Derivatives

Boronophenylalanine (BPA) is a dihydroxyboryl derivative of phenylalanine, used in neutron capture therapy.

Threonine

Threonine (abbreviated as Thr or T) encoded by the codons ACU, ACC, ACA, and ACG is an α-amino acid that is used in the biosynthesis of proteins. It contains an α-amino group (which is in the protonated −NH+3 form under biological conditions), an α-carboxylic acid group (which is in the deprotonated –COO⁻ form under biological conditions), and an alcohol containing side chain, classifying it as a polar, uncharged(at physiological pH) amino acid. It is essential in humans, meaning the body cannot synthesize it, and must be ingested in our diet. Threonine is synthesized from aspartate in bacteria such as *E. coli*.

Threonine sidechains are often hydrogen bonded; the commonest small motifs formed are ST turns, ST motifs (often at the beginning of alpha helices) and ST staples (usually at the middle of alpha helices).

Stereoisomerism

The threonine residue is susceptible to numerous posttranslational modifications. The hydroxyl side-chain can undergo *O*-linked glycosylation. In addition, threonine residues undergo phosphorylation through the action of a threonine kinase. In its phosphorylated form, it can be referred to as phosphothreonine.

It is a precursor of glycine, and can be used as a prodrug to reliably elevate brain glycine levels.

History

Threonine was discovered as the last of the 20 common proteinogenic amino acids in 1935s by William Cumming Rose, collaborating with Curtis Meyer and William Rose. The amino acid was named threonine because it was similar in structure to threose, a four-carbon monosaccharide or carbohydrate with molecular formula $C_4H_8O_4$

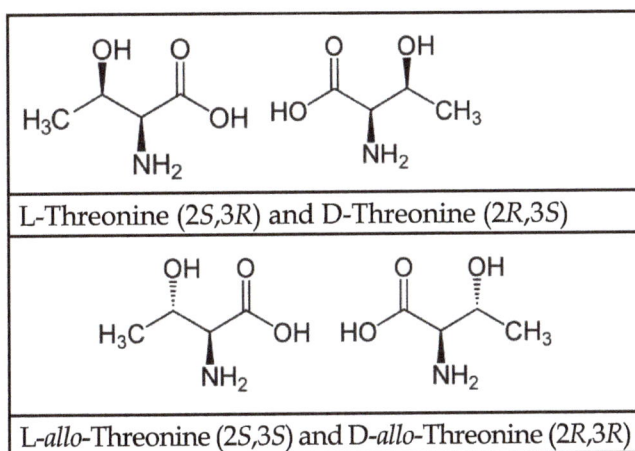

L-Threonine (2*S*,3*R*) and D-Threonine (2*R*,3*S*)

L-*allo*-Threonine (2*S*,3*S*) and D-*allo*-Threonine (2*R*,3*R*)

Threonine is one of two proteinogenic amino acids with two chiral centers. Threonine can exist in four possible stereoisomers with the following configurations: (2*S*,3*R*), (2*R*,3*S*), (2*S*,3*S*) and (2*R*,3*R*). However, the name L-threonine is used for one single diastereomer, (2*S*,3*R*)-2-amino-3-hydroxybutanoic acid. The second stereoisomer (2*S*,3*S*), which is rarely present in nature, is called L-*allo*-threonine. The two stereoisomers (2*R*,3*S*)- and (2*R*,3*R*)-2-amino-3-hydroxybutanoic acid are only of minor importance.

Biosynthesis

As an essential amino acid, threonine is not synthesized in humans, hence we must ingest threonine in the form of threonine-containing proteins. In plants and microorganisms, threonine is synthesized from aspartic acid via α-aspartyl-semialdehyde and homoserine. Homoserine undergoes *O*-phosphorylation; this phosphate ester undergoes hydrolysis concomitant with relocation of the OH group. Enzymes involved in a typical biosynthesis of threonine include:

- aspartokinase
- β-aspartate semialdehyde dehydrogenase
- homoserine dehydrogenase
- homoserine kinase
- threonine synthase.

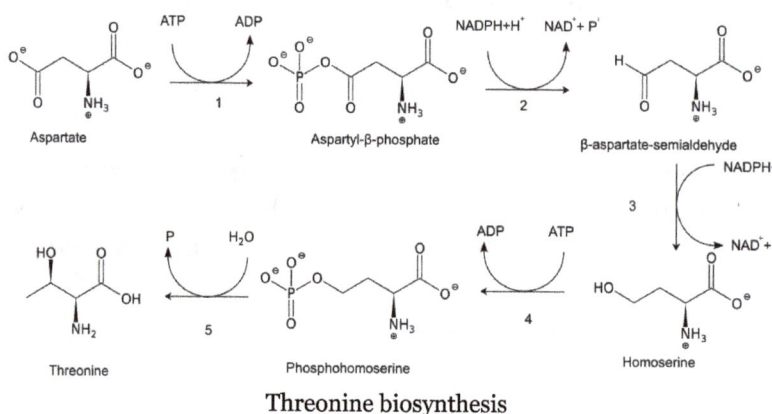

Threonine biosynthesis

Metabolism

Threonine is metabolized in two ways:

- It is converted to pyruvate via threonine dehydrogenase. An intermediate in this pathway can undergo thiolysis with CoA to produce acetyl-CoA and glycine.

- In humans, it is converted to α-ketobutyrate in a less common pathway via the enzyme serine dehydratase, and thereby enters the pathway leading to succinyl-CoA.

Sources

Foods high in threonine include cottage cheese, poultry, fish, meat, lentils, Black turtle bean and Sesame seeds.

Racemic threonine can be prepared from crotonic acid by alpha-functionalization using mercury(II) acetate.

Tryptophan

Tryptophan (abbreviated as Trp or W; encoded by the codon UGG) is an α-amino acid that is used in the biosynthesis of proteins. It contains an α-amino group (which is in the protonated $-NH_3^+$ form under biological conditions), an α-carboxylic acid group (which is in the deprotonated $-COO^-$ form under biological conditions), and a side chain indole, classifying it as a non-polar, aromatic amino acid. It is essential in humans, meaning the body cannot synthesize it and thus it must be obtained from the diet.

Tryptophan is also a precursor to neurotransmitters serotonin and melatonin.

Isolation

The isolation of tryptophan was first reported by Frederick Hopkins in 1901 through hydrolysis of casein. From 600 grams of crude casein one obtains 4-8 grams of tryptophan.

Biosynthesis and Industrial Production

As an essential amino acid, tryptophan is not synthesized from more basic substances in humans and other animals, who must ingest tryptophan or tryptophan-containing proteins. Plants and microorganisms commonly synthesize tryptophan from shikimic acid or anthranilate by the following process: anthranilate condenses with phosphoribosylpyrophosphate (PRPP), generating pyrophosphate as a by-product. The ring of the ribose moiety is opened and subjected to reductive decarboxylation, producing indole-3-glycerinephosphate; this, in turn, is transformed into indole. In the last step, tryptophan synthase catalyzes the formation of tryptophan from indole and the amino acid serine.

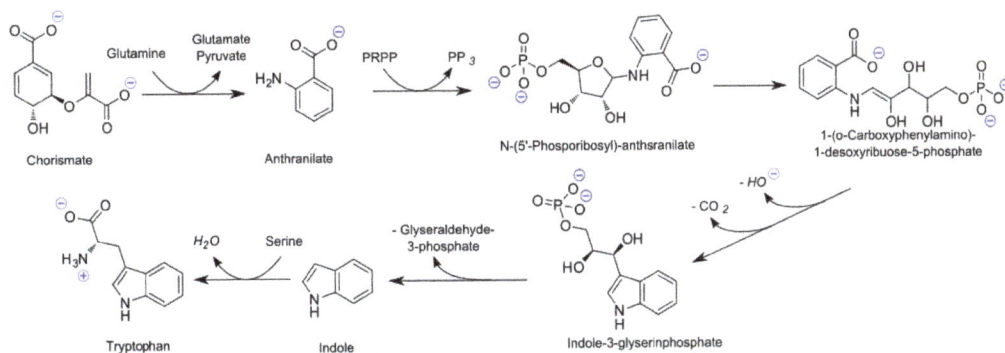

The industrial production of tryptophan is also biosynthetic and is based on the fermentation of serine and indole using either wild-type or genetically modified bacteria such as *B. amyloliquefaciens*, *B. subtilis*, *C. glutamicum* or *E. coli*. These strains carry either mutations that prevent the reuptake of aromatic amino acids or multiple/overexpressed trp operons. The conversion is catalyzed by the enzyme tryptophan synthase.

Function

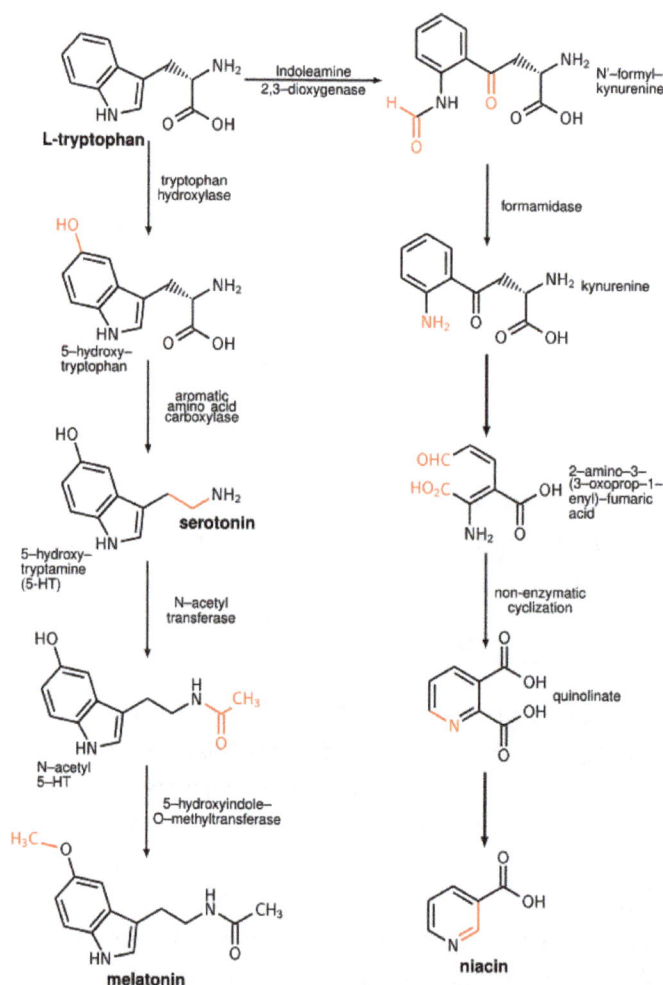

Metabolism of L-tryptophan into serotonin and melatonin (left) and niacin (right). Transformed functional groups after each chemical reaction are highlighted in red.

For many organisms (including humans), tryptophan is needed to prevent illness or death, but cannot be synthesized by the organism and must be ingested; in short, it is an essential amino acid. Amino acids, including tryptophan, act as building blocks in protein biosynthesis, and proteins are required to sustain life. In addition, tryptophan functions as a biochemical precursor for the following compounds:

- Serotonin (a neurotransmitter), synthesized via tryptophan hydroxylase. Serotonin, in turn, can be converted to melatonin (a neurohormone), via N-acetyl-transferase and 5-hydroxyindole-O-methyltransferase activities.

- Niacin, also known as vitamin B_3, is synthesized from tryptophan via kynurenine and quinolinic acids as key biosynthetic intermediates.

- Auxins (a class of phytohormones) are synthesized from tryptophan.

The disorder fructose malabsorption causes improper absorption of tryptophan in the intestine, reduced levels of tryptophan in the blood, and depression. Some studies did not find reduced tryptophan in cases of lactose maldigestion.

In bacteria that synthesize tryptophan, high cellular levels of this amino acid activate a repressor protein, which binds to the trp operon. Binding of this repressor to the tryptophan operon prevents transcription of downstream DNA that codes for the enzymes involved in the biosynthesis of tryptophan. So high levels of tryptophan prevent tryptophan synthesis through a negative feedback loop and, when the cell's tryptophan levels are reduced, transcription from the trp operon resumes. The genetic organisation of the trp operon thus permits tightly regulated and rapid responses to changes in the cell's internal and external tryptophan levels.

Tryptophan Metabolism by Human Gastrointestinal Microbiota (v · t · e)

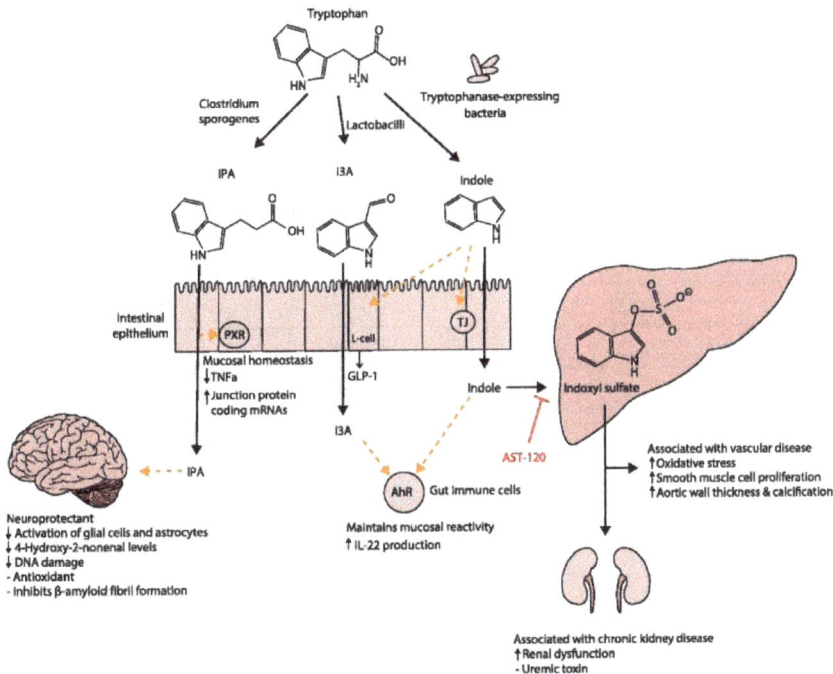

This diagram shows the biosynthesis of bioactive compounds (indole and certain derivatives) from tryptophan by bacteria in the gut. Indole is produced from tryptophan by bacteria that express tryptophanase. *Clostridium sporogenes* metabolizes indole into 3-indolepropionic acid (IPA), a highly potent neuroprotective antioxidant that scavenges hydroxyl radicals. In the intestine, IPA binds to pregnane X receptors (PXR) in intestinal cells, thereby facilitating mucosal homeostasis and barrier function. Following absorption from the intestine and distribution to the brain, IPA confers a neuroprotective effect against cerebral ischemia and Alzheimer's disease. *Lactobacillus* species metabolize tryptophan into indole-3-aldehyde (I3A) which acts on the aryl hydrocarbon receptor (AhR) in intestinal immune cells, in turn increasing interleukin-22 (IL-22) production. Indole itself acts as a glucagon-like peptide-1 (GLP-1) secretagogue in intestinal L cells and as a ligand for AhR. Indole can also be metabolized by the liver into indoxyl sulfate, a compound that is toxic in high concentrations and associated with vascular disease and renal dysfunction. AST-120 (activated charcoal), an intestinal sorbent that is taken by mouth, adsorbs indole, in turn decreasing the concentration of indoxyl sulfate in blood plasma.

Dietary Sources

Tryptophan is a routine constituent of most protein-based foods or dietary proteins. It is particularly plentiful in chocolate, oats, dried dates, milk, yogurt, cottage cheese, red meat, eggs, fish, poultry, sesame, chickpeas, almonds, sunflower seeds, pumpkin seeds, buckwheat, spirulina, bananas, and peanuts. Contrary to the popular belief that turkey contains an abundance of tryptophan, the tryptophan content in turkey is typical of poultry.

Tryptophan (Trp) Content of Various Foods			
Food	Tryptophan [g/100 g of food]	Protein [g/100 g of food]	Tryptophan/Protein [%]
egg white, dried	1.00	81.10	1.23
spirulina, dried	0.93	57.47	1.62
cod, atlantic, dried	0.70	62.82	1.11
soybeans, raw	0.59	36.49	1.62
cheese, Parmesan	0.56	37.90	1.47
sesame seed	0.37	17.00	2.17
cheese, cheddar	0.32	24.90	1.29
sunflower seed	0.30	17.20	1.74
pork, chop	0.25	19.27	1.27
turkey	0.24	21.89	1.11
chicken	0.24	20.85	1.14
beef	0.23	20.13	1.12
oats	0.23	16.89	1.39
salmon	0.22	19.84	1.12
lamb, chop	0.21	18.33	1.17
perch, Atlantic	0.21	18.62	1.12
chickpeas, raw	0.19	19.30	0.96
egg	0.17	12.58	1.33
wheat flour, white	0.13	10.33	1.23
baking chocolate, unsweetened	0.13	12.9	1.23
milk	0.08	3.22	2.34
rice, white, medium-grain, cooked	0.028	2.38	1.18
quinoa, uncooked	0.167	14.12	1.2
quinoa, cooked	0.052	4.40	1.1
potatoes, russet	0.02	2.14	0.84
tamarind	0.018	2.80	0.64
banana	0.01	1.03	0.87

Turkey Meat and Drowsiness

A common assertion in the US is that heavy consumption of turkey meat results in drowsiness, due to high levels of tryptophan contained in turkey. However, the amount of tryptophan in turkey is comparable to that contained in other meats. Drowsiness after eating may be caused by other foods eaten with the turkey, particularly carbohydrates. It has been demonstrated in both animal and human tests that ingestion of a meal rich in carbohydrates triggers release of insulin. Insulin in turn stimulates the uptake of large neutral branched-chain amino acids (BCAA), but not tryptophan into muscle, increasing the ratio of tryptophan to BCAA in the blood stream. The resulting increased tryptophan ratio reduces competition at the large neutral amino acid transporter (which transports both BCAA and aromatic amino acids), resulting in more uptake of tryptophan across the blood–brain barrier into the cerebrospinal fluid (CSF). Once in the CSF, tryptophan is converted into serotonin in the raphe nuclei by the normal enzymatic pathway. The resultant serotonin is further metabolised into melatonin by the pineal gland. Hence, this data suggests that "feast-induced drowsiness"—or postprandial somnolence—may be the result of a heavy meal rich in carbohydrates, which indirectly increases the production of sleep-promoting melatonin in the brain.

Use as a Dietary Supplement

Tryptophan is sold over the counter in the United States, Canada, and the United Kingdom as a dietary supplement for use as an antidepressant, anxiolytic, and sleep aid. It is also marketed as a prescription drug in some European countries for the indication of major depression under various trade names.

Since tryptophan is converted into 5-hydroxytryptophan (5-HTP) which is subsequently converted into the neurotransmitter serotonin, it has been proposed that consumption of tryptophan or 5-HTP may therefore improve depression symptoms by increasing the level of serotonin in the brain. In 2001 a Cochrane Review of the effect of 5-HTP and tryptophan on depression was published. The authors included only studies of a high rigor and included both 5-HTP and tryptophan in their review because of the limited data on either. Of 108 studies of 5-HTP and tryptophan on depression published between 1966 and 2000, only two met the authors' quality standards for inclusion, totaling 64 study participants. The substances were more effective than placebo in the two studies included but the authors state that, "the evidence was of insufficient quality to be conclusive," and note, "because alternative antidepressants exist which have been proven to be effective and safe, the clinical usefulness of 5-HTP and tryptophan is limited at present." The use of tryptophan as an adjunctive therapy in addition to standard treatment for mood and anxiety disorders is not supported by the scientific evidence. Due to the lack of high quality studies and preliminary nature of studies showing effectiveness and the lack of adequate study on their safety, the use of tryptophan and 5-HTP is not highly recommended or thought to be clinically useful.

There is evidence that blood tryptophan levels are unlikely to be altered by changing the diet, but tryptophan is available in health food stores as a dietary supplement. Consuming purified tryptophan increases brain serotonin whereas eating foods containing tryptophan does not. This is because the transport system which brings tryptophan across the blood-brain barrier is also selective for the other amino acids which are contained in protein food sources. High blood plasma levels of other large neutral amino acids prevent the plasma concentration of tryptophan from increasing brain concentration levels.

Side Effects

Potential side effects of tryptophan include nausea, diarrhea, drowsiness, lightheadedness, headache, dry mouth, blurred vision, sedation, euphoria, and nystagmus (involuntary eye movements). Because tryptophan has not been thoroughly studied in a clinical setting, possible side effects and interactions with other drugs are not well known. However, while use of this amino acid to treat depression or mood by elevating serotonin levels may or may not be effective in some individuals (in some cases, in some situations, for an indeterminate period of time), as with any shift in neurochemical balances, altering a person's neurochemistry is never a simple prospect and may give rise to mental and emotional instabilities, a worsening of depression or mood (potentially as far as suicidal thoughts or impulses), or other detrimental effects therapeutically inadvisable.

Interactions

Tryptophan taken as a dietary supplement (such as in tablet form) has the potential to cause serotonin syndrome when combined with antidepressants of the MAOI or SSRI class or other strongly serotonergic drugs. Normal dietary intake of tryptophan in food would not be expected to cause serotonin syndrome.

Research

In 1912 Felix Ehrlich demonstrated that yeast attacks the natural amino acids essentially by splitting off carbon dioxide and replacing the amino group with hydroxyl. By this reaction, tryptophan gives rise to tryptophol.

Tryptophan affects brain serotonin synthesis when given orally in a purified form and is used to modify serotonin levels for research in psychology. Low brain serotonin is induced by administration of tryptophan-poor protein in a technique called 'acute tryptophan depletion'. Studies using this method have evaluated the effect of serotonin on mood and social behavior, finding that serotonin reduces aggression and increases agreeableness.

Fluorescence

Tryptophan is an important intrinsic fluorescent probe (amino acid), which can be used to estimate the nature of microenvironment of the tryptophan. Most of the intrinsic fluorescence emissions of a folded protein are due to excitation of tryptophan residues.

Safety

Eosinophilia–myalgia Syndrome

There was a large outbreak of eosinophilia-myalgia syndrome (EMS) in the U.S. in 1989, with more than 1,500 cases reported to the CDC and at least 37 deaths. After preliminary investigation revealed that the outbreak was linked to intake of tryptophan, the U.S. Food and Drug Administration (FDA) banned most tryptophan from sale in the US in 1991, and other countries followed suit.

Subsequent epidemiological studies suggested that EMS was linked to specific batches of L-tryptophan supplied by a single large Japanese manufacturer, Showa Denko. It eventually became clear that recent batches of Showa Denko's L-tryptophan were contaminated by trace impurities, which were subsequently thought to be responsible for the 1989 EMS outbreak. However, other evidence suggests that tryptophan itself may be a potentially major contributory factor in EMS.

The FDA loosened its restrictions on sales and marketing of tryptophan in February 2001, but continued to limit the importation of tryptophan not intended for an exempted use until 2005.

The fact that the Showa Denko facility used genetically engineered bacteria to produce the contaminated batches of L-tryptophan later found to have caused the outbreak of eosinophilia-myalgia syndrome has been cited as evidence of a need for "close monitoring of the chemical purity of biotechnology-derived products." Those calling for purity monitoring have, in turn, been criticized as anti-GMO activists who overlook possible non-GMO causes of contamination and threaten the development of biotech.

Valine

Valine (abbreviated as Val or V) encoded by the codons GUU, GUC, GUA, and GUG is an α-amino acid that is used in the biosynthesis of proteins. It contains an α-amino group (which is in the protonated $-NH_3^+$ form under biological conditions), an α-carboxylic acid group (which is in the deprotonated $-COO^-$ form under biological conditions), and a side chain isopropyl variable group, classifying it as a non-polar amino acid. It is essential in humans, meaning the body cannot synthesize it and thus it must be obtained from the diet. Human dietary sources are any proteinaceous foods such as meats, dairy products, soy products, beans and legumes.

Along with leucine and isoleucine, valine is a branched-chain amino acid. In sickle-cell disease, valine substitutes for the hydrophilic amino acid glutamic acid in β-globin. Because valine is hydrophobic, the hemoglobin is prone to abnormal aggregation.

History and Etymology

Valine was first isolated from casein in 1901 by Hermann Emil Fischer. The name valine comes from valeric acid, which in turn is named after the plant valerian due to the presence of the acid in the roots of the plant.

Nomenclature

According to IUPAC, carbon atoms forming valine are numbered sequentially starting from 1 denoting the carboxyl carbon, whereas 4 and 4' denote the two terminal methyl carbons.

Biosynthesis

Valine is an essential amino acid, hence it must be ingested, usually as a component of proteins. It is synthesized in plants via several steps starting from pyruvic acid. The initial part of the pathway also leads to leucine. The intermediate α-ketoisovalerate undergoes reductive amination with glutamate. Enzymes involved in this biosynthesis include:

- Acetolactate synthase (also known as acetohydroxy acid synthase)
- Acetohydroxy acid isomeroreductase
- Dihydroxyacid dehydratase
- Valine aminotransferase

Synthesis

Racemic valine can be synthesized by bromination of isovaleric acid followed by amination of the α-bromo derivative

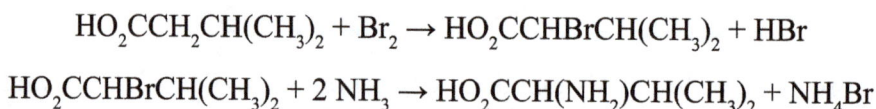

$$HO_2CCH_2CH(CH_3)_2 + Br_2 \rightarrow HO_2CCHBrCH(CH_3)_2 + HBr$$

$$HO_2CCHBrCH(CH_3)_2 + 2\ NH_3 \rightarrow HO_2CCH(NH_2)CH(CH_3)_2 + NH_4Br$$

Valine and Insulin Resistance

Valine, as well as other branched-chain amino acids, are associated with insulin resistance, as higher levels of valine are observed in the blood of diabetic mice, rats, and humans. Mice fed a valine free diet for one day have improved insulin sensitivity, and feeding of a valine free diet for one week significantly decreases blood glucose levels. The valine catabolite 3-hydroxyisobutyrate promotes skeletal muscle insulin resistance in mice by stimulating fatty acid uptake into muscle and lipid accumulation. In humans, a protein restricted diet lowers blood levels of valine and decreases fasting blood glucose levels.

References

- Roberts, John D. (2000). ABCs of FT-NMR. Sausalito, CA: University Science Books. pp. 258–9. ISBN 978-1-891389-18-4.

- Nelson, D. L.; Cox, M. M. "Lehninger, Principles of Biochemistry" 3rd Ed. Worth Publishing: New York, 2000. ISBN 1-57259-153-6.

- Winter, Ruth (2009). A consumer's dictionary of food additives (7th ed.). New York: Three Rivers Press. ISBN 0307408922.

- Connor, J.M.; "Global Price Fixing" 2nd Ed. Springer-Verlag: Heidelberg, 2008. ISBN 978-3-540-78669-6.

- Eichenwald, Kurt.; "The Informant: a true story" Broadway Books: New York, 2000. ISBN 0-7679-0326-9.

- Weast, Robert C., ed. (1981). CRC Handbook of Chemistry and Physics (62nd ed.). Boca Raton, FL: CRC Press. p. C-374. ISBN 0-8493-0462-8. .

- Harvey M. Ross; June Roth (1 April 1991). The Mood Control Diet: 21 Days to Conquering Depression and Fatigue. Simon & Schuster. p. 59. ISBN 978-0-13-590449-7.

- Nelson, D. L.; Cox, M. M. (2000). Lehninger, Principles of Biochemistry (3rd ed.). New York: Worth Publishing. ISBN 1-57259-153-6.

- Sprenger, G. A. (2007). "Aromatic Amino Acids". Amino Acid Biosynthesis: Pathways, Regulation and Metabolic Engineering (1st ed.). Springer. pp. 106–113. ISBN 978-3-540-48595-7.

- Lehninger, Albert L.; Nelson, David L.; Cox, Michael M. (2000), Principles of Biochemistry (3rd ed.), New York: W. H. Freeman, ISBN 1-57259-153-6 .

- Dawson, RMC; et al. (1969). Data for Biochemical Research. Oxford: Clarendon Press. ISBN 0-19-855338-2.

- Weast, Robert C., ed. (1981). CRC Handbook of Chemistry and Physics (62nd ed.). Boca Raton, FL: CRC Press. p. C-569. ISBN 0-8493-0462-8.

- Jones, J. H., ed. (1985). Amino Acids, Peptides and Proteins. Specialist Periodical Reports. 16. London: Royal Society of Chemistry. p. 389. ISBN 978-0-85186-144-9.

- Lehninger, Albert L.; Nelson, David L.; Cox, Michael M. (2000), Principles of Biochemistry (3rd ed.), New York: W. H. Freeman, ISBN 1-57259-153-6 .

Various Types of Amino Acid

Acids which are very important for metabolism are found abundantly in nature and have to be consumed for bodily utilization. The various types of amino acids discussed are glutamic acid, proteinogenic amino acid, non-proteinogenic amoino acids, beta-Alanie, gamma-aminobutyric acid, citulline etc. The major categories of amino acids are dealt with great details in this section.

Glutamic Acid

Glutamic acid (abbreviated as Glu or E; encoded by the codons GAA or GAG) is an α-amino acid that is used in the biosynthesis of proteins. It contains an α-amino group (which is in the protonated $-NH_3^+$ form under biological conditions), an α-carboxylic acid group (which is in the deprotonated $-COO^-$ form under biological conditions), and a side chain carboxylic acid, classifying it as a polar negatively charged (at physiological pH), aliphatic amino acid. It is non-essential in humans, meaning the body can synthesize it.

In neuroscience, its carboxylate anion glutamate is an important excitatory neurotransmitter that plays the principal role in neural activation.

The conjugate base is glutamate; the radical is glutamyl.

Chemistry

The side chain carboxylic acid functional group has a pK_a of 4.1 and therefore exists almost entirely in its negatively charged deprotonated carboxylate form at pH values greater than 4.1; therefore, it is negatively charged at physiological pH ranging from 7.35 to 7.45.

History

Although they occur naturally in many foods, the flavor contributions made by glutamic acid and other amino acids were only scientifically identified early in the twentieth century. The substance was discovered and identified in the year 1866, by the German chemist Karl Heinrich Ritthausen who treated wheat gluten (for which it was named) with sulfuric acid. In 1908 Japanese researcher Kikunae Ikeda of the Tokyo Imperial University identified brown crystals left behind after the evaporation of a large amount of kombu broth as glutamic acid. These crystals, when tasted, reproduced the ineffable but undeniable flavor he detected in many foods, most especially in seaweed. Professor Ikeda termed this flavor umami. He then patented a method of mass-producing a crystalline salt of glutamic acid, monosodium glutamate.

Biosynthesis

Reactants	Products	Enzymes
Glutamine + H_2O	$\rightarrow$ Glu + NH_3	GLS, GLS2
NAcGlu + H_2O	$\rightarrow$ Glu + Acetate	N-acetyl-glutamate synthase
α-ketoglutarate + NADPH + NH_4^+	$\rightarrow$ Glu + $NADP^+$ + H_2O	GLUD1, GLUD2
α-ketoglutarate + α-amino acid	$\rightarrow$ Glu + α-keto acid	transaminase
1-Pyrroline-5-carboxylate + NAD^+ + H_2O	$\rightarrow$ Glu + NADH	ALDH4A1
N-formimino-L-glutamate + FH_4	$\rightarrow$ Glu + 5-formimino-FH_4	FTCD
NAAG	$\rightarrow$ Glu + NAA	GCPII

Function and Uses

Metabolism

Glutamate is a key compound in cellular metabolism. In humans, dietary proteins are broken down by digestion into amino acids, which serve as metabolic fuel for other functional roles in the body. A key process in amino acid degradation is transamination, in which the amino group of an amino acid is transferred to an α-ketoacid, typically catalysed by a transaminase. The reaction can be generalised as such:

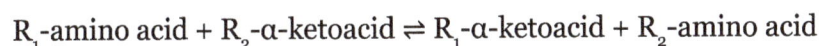

R_1-amino acid + R_2-α-ketoacid $\rightleftharpoons$ R_1-α-ketoacid + R_2-amino acid

A very common α-keto acid is α-ketoglutarate, an intermediate in the citric acid cycle. Transamination of α-ketoglutarate gives glutamate. The resulting α-ketoacid product is often a useful one as well, which can contribute as fuel or as a substrate for further metabolism processes. Examples are as follows:

Alanine + α-ketoglutarate $\rightleftharpoons$ pyruvate + glutamate

Aspartate + α-ketoglutarate $\rightleftharpoons$ oxaloacetate + glutamate

Both pyruvate and oxaloacetate are key components of cellular metabolism, contributing as substrates or intermediates in fundamental processes such as glycolysis, gluconeogenesis, and the citric acid cycle.

Glutamate also plays an important role in the body's disposal of excess or waste nitrogen. Glutamate undergoes deamination, an oxidative reaction catalysed by glutamate dehydrogenase, as follows:

glutamate + H_2O + $NADP^+$ $\rightarrow$ α-ketoglutarate + NADPH + NH_3 + H^+

Ammonia (as ammonium) is then excreted predominantly as urea, synthesised in the liver. Transamination can thus be linked to deamination, effectively allowing nitrogen from the amine groups of amino acids to be removed, via glutamate as an intermediate, and finally excreted from the body in the form of urea.

Glutamate is also a neurotransmitter (see below), which makes it one of the most abundant molecules in the brain. Malignant brain tumors known as glioma or glioblastoma exploit this phenomenon by using glutamate as an energy source, especially when these mutations become more dependent on glutamate due to mutations in the gene *IDH1*.

Neurotransmitter

Glutamate is the most abundant excitatory neurotransmitter in the vertebrate nervous system. At chemical synapses, glutamate is stored in vesicles. Nerve impulses trigger release of glutamate from the presynaptic cell. Glutamate acts on ionotropic and metabotropic (G-protein coupled) receptors. In the opposing postsynaptic cell, glutamate receptors, such as the NMDA receptor or the AMPA receptor, bind glutamate and are activated. Because of its role in synaptic plasticity, glutamate is involved in cognitive functions such as learning and memory in the brain. The form of plasticity known as long-term potentiation takes place at glutamatergic synapses in the hippocampus, neocortex, and other parts of the brain. Glutamate works not only as a point-to-point transmitter, but also through spill-over synaptic crosstalk between synapses in which summation of glutamate released from a neighboring synapse creates extrasynaptic signaling/volume transmission. In addition, glutamate plays important roles in the regulation of growth cones and synaptogenesis during brain development as originally described by Mark Mattson.

Brain Nonsynaptic Glutamatergic Signaling Circuits

Extracellular glutamate in *Drosophila* brains has been found to regulate postsynaptic glutamate receptor clustering, via a process involving receptor desensitization. A gene expressed in glial cells actively transports glutamate into the extracellular space, while, in the nucleus accumbens-stimulating group II metabotropic glutamate receptors, this gene was found to reduce extracellular glutamate levels. This raises the possibility that this extracellular glutamate plays an "endocrine-like" role as part of a larger homeostatic system.

GABA Precursor

Glutamate also serves as the precursor for the synthesis of the inhibitory gamma-aminobutyric acid (GABA) in GABA-ergic neurons. This reaction is catalyzed by glutamate decarboxylase (GAD), which is most abundant in the cerebellum and pancreas.

Stiff person syndrome is a neurologic disorder caused by anti-GAD antibodies, leading to a decrease in GABA synthesis and, therefore, impaired motor function such as muscle stiffness and spasm. Since the pancreas has abundant GAD, a direct immunological destruction occurs in the pancreas and the patients will have diabetes mellitus.

Flavor Enhancer

Glutamic acid, being a constituent of protein, is present in every food that contains protein, but it can only be tasted when it is present in an unbound form. Significant amounts of free glutamic acid are present in a wide variety of foods, including cheese and soy sauce, and is responsible for umami, one of the five basic tastes of the human sense of taste. Glutamic acid is often used as a food additive and flavor enhancer in the form of its salt, known as monosodium glutamate (MSG).

Nutrient

All meats, poultry, fish, eggs, dairy products, and kombu are excellent sources of glutamic acid. Some protein-rich plant foods also serve as sources. 30% to 35% of the protein in wheat is glutamic acid. Ninety-five percent of the dietary glutamate is metabolized by intestinal cells in a first pass.

Plant Growth

Auxigro is a plant growth preparation that contains 30% glutamic acid.

NMR Spectroscopy

In recent years, there has been much research into the use of residual dipolar coupling (RDC) in nuclear magnetic resonance spectroscopy (NMR). A glutamic acid derivative, poly-γ-benzyl-L-glutamate (PBLG), is often used as an alignment medium to control the scale of the dipolar interactions observed.

Pharmacology

The drug phencyclidine (more commonly known as PCP) antagonizes glutamic acid non-competitively at the NMDA receptor. For the same reasons, dextromethorphan and ketamine also have strong dissociative and hallucinogenic effects. Acute infusion of the drug LY354740 (also known as eglumegad, an agonist of the metabotropic glutamate receptors 2 and 3) resulted in a marked diminution of yohimbine-induced stress response in bonnet macaques (Macaca radiata); chronic oral administration of LY354740 in those animals led to markedly reduced baseline cortisol levels (approximately 50 percent) in comparison to untreated control subjects. LY354740 has also been demonstrated to act on the metabotropic glutamate receptor 3 (GRM3) of human adrenocortical cells, downregulating aldosterone synthase, CYP11B1, and the production of adrenal steroids (i.e. aldosterone and cortisol). Glutamate does not easily pass the blood brain barrier, but, instead, is transported by a high-affinity transport system. It can also be converted into glutamine.

Proteinogenic Amino Acid

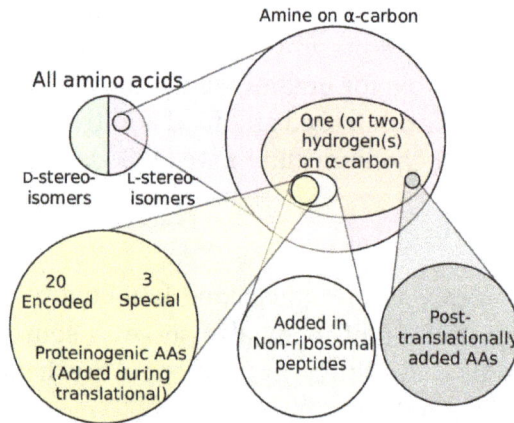

Proteinogenic amino acids are a small fraction of all amino acids

Proteinogenic amino acids are amino acids that are precursors to proteins, and are incorporated into proteins during translation. Throughout known life, there are 23 proteinogenic amino acids, 20 in the standard genetic code and an additional 3 that can be incorporated by special translation mechanisms.

Both eukaryotes and prokaryotes can incorporate selenocysteine into their proteins via a nucleotide sequence known as a SECIS element, which directs the cell to translate a nearby UGA codon as selenocysteine (UGA is normally a stop codon). In some methanogenic prokaryotes, the UAG codon (normally a stop codon) can also be translated to pyrrolysine. In bacteria, the AUG initiation codon is translated to N-formylmethionine when it is actually used to initiate translation and translated normally (to methionine) at other times.

In eukaryotes, there are only 21 proteinogenic amino acids, the 20 of the standard genetic code, plus selenocysteine. Humans can synthesize 12 of these from each other or from other molecules of intermediary metabolism. The other nine must be consumed (usually as their protein derivatives), and so they are called essential amino acids. The essential amino acids are histidine, isoleucine, leucine, lysine, methionine, phenylalanine, threonine, tryptophan, and valine (i.e. H, I, L, K, M, F, T, W, V).

The word "proteinogenic" means "protein creating". Proteinogenic amino acids can be condensed into a polypeptide (the subunit of a protein) through a process called translation (the second stage of protein biosynthesis, part of the overall process of gene expression).

In contrast, non-proteinogenic amino acids are either not incorporated in proteins (like GABA, L-DOPA, or triiodothyronine), or are not produced directly and in isolation by standard cellular machinery (like hydroxyproline and selenomethionine). The latter often results from post-translational modification of proteins.

The proteinogenic amino acids have been found to be related to the set of amino acids that can be recognized by ribozyme autoaminoacylation systems. Thus, nonproteinogenic amino acids would have been excluded by the contingent evolutionary success of nucleotide-based life forms. Other reasons have been offered to explain why certain specific nonproteinogenic amino acids are not generally incorporated into proteins; for example, ornithine and homoserine cyclize against the peptide backbone and fragment the protein with relatively short half-lives, while others are toxic because they can be mistakenly incorporated into proteins, such as the arginine analog canavanine.

Nonproteinogenic amino acids are incorporated in nonribosomal peptides, which are not produced by the ribosome during translation.

Structures

The following illustrates the structures and abbreviations of the 21 amino acids that are directly encoded for protein synthesis by the genetic code of eukaryotes. The structures given below are standard chemical structures, not the typical zwitterion forms that exist in aqueous solutions.

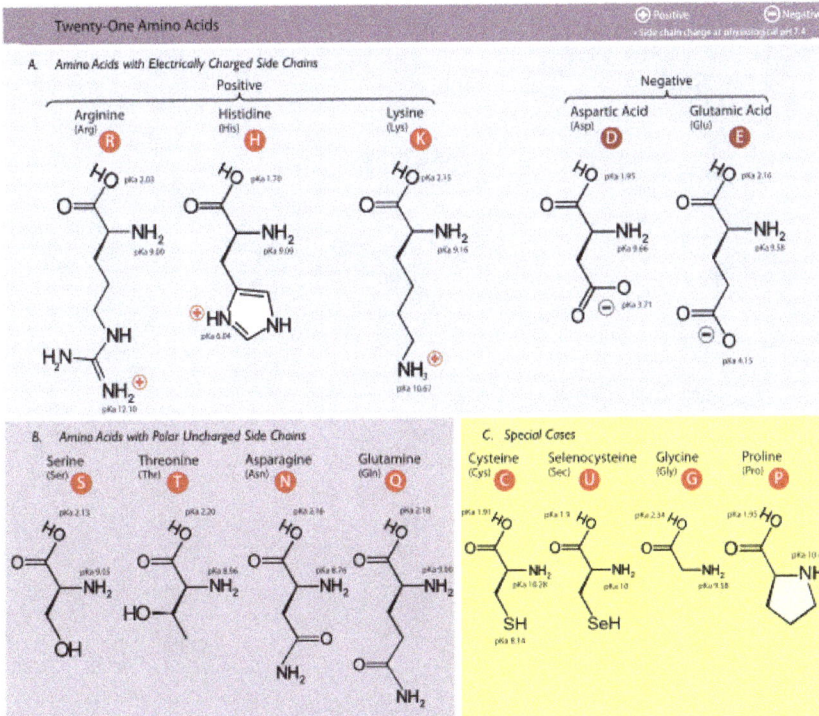

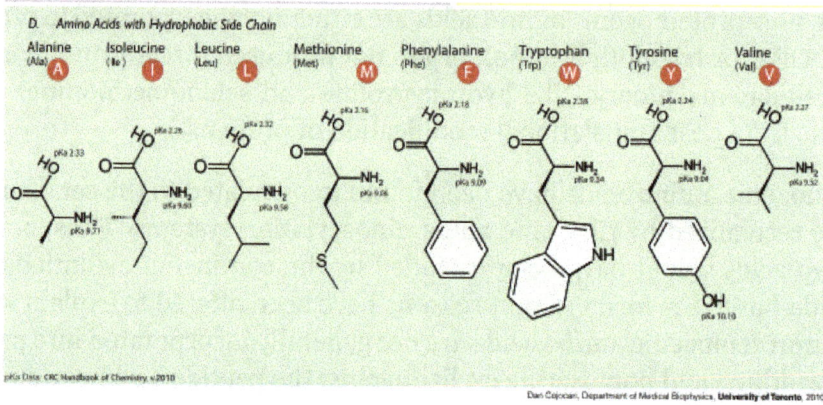

D. Amino Acids with Hydrophobic Side Chain

Alanine (Ala) A | Isoleucine (Ile) I | Leucine (Leu) L | Methionine (Met) M | Phenylalanine (Phe) F | Tryptophan (Trp) W | Tyrosine (Tyr) Y | Valine (Val) V

pKa Data: CRC Handbook of Chemistry, v.2010

Dan Cojocari, Department of Medical Biophysics, University of Toronto, 2010

Grouped table of 21 amino acids' structures, nomenclature, and their side groups' pKa values

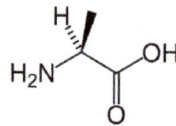

L-Alanine
(Ala / A)

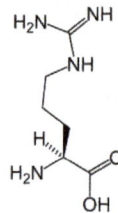

L-Arginine
(Arg / R)

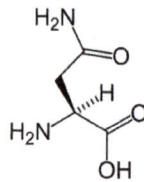

L-Asparagine
(Asn / N)

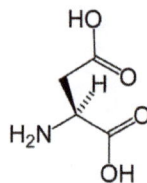

L-Aspartic acid
(Asp / D)

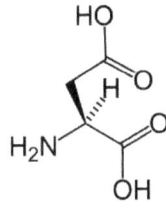

L-Cysteine
(Cys / C)

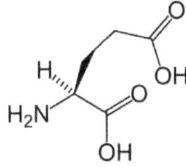

L-Glutamic acid
(Glu / E)

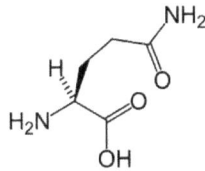

L-Glutamine
(Gln / Q)

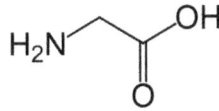

Glycine
(Gly / G)

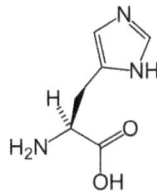

L-Histidine
(His / H)

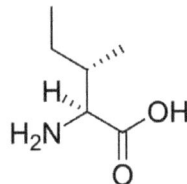

L-Isoleucine
(Ile / I)

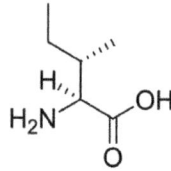

L-Leucine
(Leu / L)

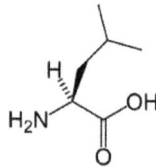

L-Lysine
(Lys / K)

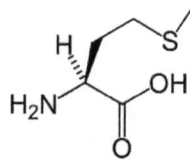

L-Methionine
(Met / M)

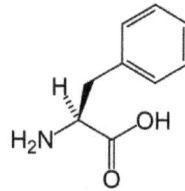

L-Phenylalanine
(Phe / F)

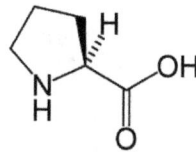

L-Proline
(Pro / P)

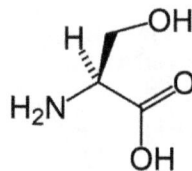

L-Serine
(Ser / S)

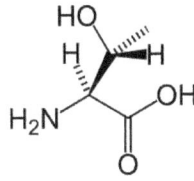

L-Threonine
(Thr / T)

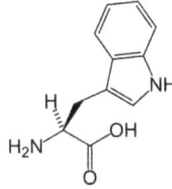

L-Tryptophan
(Trp / W)

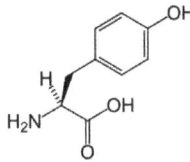

L-Tyrosine
(Tyr / Y)

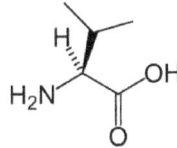

L-Valine
(Val / V)

IUPAC/IUBMB now also recommends standard abbreviations for the following two amino acids:

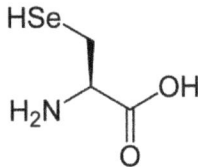

L-Selenocysteine
(Sec / U)

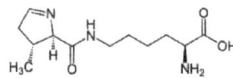

L-Pyrrolysine
(Pyl / O)

Nonspecific Abbreviations

Sometimes, the specific identity of an amino acid cannot be determined unambiguously. Certain protein sequencing techniques do not distinguish among certain pairs. Thus, these codes are used:

- Asx (B) is "asparagine or aspartic acid"

- Glx (Z) is "glutamic acid or glutamine"

- Xle (J) is "leucine or isoleucine"

In addition, the symbol X is used to indicate an amino acid that is completely unidentified.

Chemical Properties

Following is a table listing the one-letter symbols, the three-letter symbols, and the chemical properties of the side chains of the standard amino acids. The masses listed are based on weighted averages of the elemental isotopes at their natural abundances. Forming a peptide bond results in elimination of a molecule of water, so the mass of an amino acid unit within a protein chain is reduced by 18.01524 Da.

General chemical properties

Amino acid	Short	Abbrev.	Avg. mass (Da)	pI	pK_1 (α-COOH)	pK_2 (α-$^+NH_3$)
Alanine	A	Ala	89.09404	6.01	2.35	9.87
Cysteine	C	Cys	121.15404	5.05	1.92	10.70
Aspartic acid	D	Asp	133.10384	2.85	1.99	9.90
Glutamic acid	E	Glu	147.13074	3.15	2.10	9.47
Phenylalanine	F	Phe	165.19184	5.49	2.20	9.31
Glycine	G	Gly	75.06714	6.06	2.35	9.78
Histidine	H	His	155.15634	7.60	1.80	9.33
Isoleucine	I	Ile	131.17464	6.05	2.32	9.76
Lysine	K	Lys	146.18934	9.60	2.16	9.06
Leucine	L	Leu	131.17464	6.01	2.33	9.74
Methionine	M	Met	149.20784	5.74	2.13	9.28
Asparagine	N	Asn	132.11904	5.41	2.14	8.72
Pyrrolysine	O	Pyl	255.31			
Proline	P	Pro	115.13194	6.30	1.95	10.64
Glutamine	Q	Gln	146.14594	5.65	2.17	9.13
Arginine	R	Arg	174.20274	10.76	1.82	8.99
Serine	S	Ser	105.09344	5.68	2.19	9.21
Threonine	T	Thr	119.12034	5.60	2.09	9.10

Amino acid	Short	Abbrev.	Avg. mass (Da)	pI	pK$_1$ (α-COOH)	pK$_2$ (α-$^+$NH$_3$)
Selenocysteine	U	Sec	168.053	5.47	1.91	10
Valine	V	Val	117.14784	6.00	2.39	9.74
Tryptophan	W	Trp	204.22844	5.89	2.46	9.41
Tyrosine	Y	Tyr	181.19124	5.64	2.20	9.21

Side Chain Properties

Amino acid	Short	Abbrev.	Side chain	Hydro-phobic	pKa	Polar	pH	Small	Tiny	Aromatic or Aliphatic	van der Waals volume
Alanine	A	Ala	-CH$_3$	X	-	-	-	X	X	-	67
Cysteine	C	Cys	-CH$_2$SH	X	8.18	-	acidic	X	X	-	86
Aspartic acid	D	Asp	-CH$_2$COOH	-	3.90	X	acidic	X	-	-	91
Glutamic acid	E	Glu	-CH$_2$CH$_2$COOH	-	4.07	X	acidic	-	-	-	109
Phenylalanine	F	Phe	-CH$_2$C$_6$H$_5$	X	-	-	-	-	-	Aromatic	135
Glycine	G	Gly	-H	X	-	-	-	X	X	-	48
Histidine	H	His	-CH$_2$-C$_3$H$_3$N$_2$	-	6.04	X	weak basic	-	-	Aromatic	118
Isoleucine	I	Ile	-CH(CH$_3$)CH$_2$CH$_3$	X	-	-	-	-	-	Aliphatic	124
Lysine	K	Lys	-(CH$_2$)$_4$NH$_2$	-	10.54	X	basic	-	-	-	135
Leucine	L	Leu	-CH$_2$CH(CH$_3$)$_2$	X	-	-	-	-	-	Aliphatic	124
Methionine	M	Met	-CH$_2$CH$_2$SCH$_3$	X	-	-	-	-	-	-	124
Asparagine	N	Asn	-CH$_2$CONH$_2$	-	-	X	-	X	-	-	96
Pyrrolysine	O	Pyl	-(CH$_2$)$_4$NHCOC$_4$H$_5$NCH$_3$	-	-	X	weak basic	-	-	-	
Proline	P	Pro	-CH$_2$CH$_2$CH$_2$-	X	-	-	-	X	-	-	90
Glutamine	Q	Gln	-CH$_2$CH$_2$CONH$_2$	-	-	X	weak basic	-	-	-	114
Arginine	R	Arg	-(CH$_2$)$_3$NH-C(NH)NH$_2$	-	12.48	X	strongly basic	-	-	-	148
Serine	S	Ser	-CH$_2$OH	-	-	X	weak acidic	X	X	-	73
Threonine	T	Thr	-CH(OH)CH$_3$	-	-	X	weak acidic	X	-	-	93
Selenocysteine	U	Sec	-CH$_2$SeH	-	5.73	-	acidic	X	X	-	
Valine	V	Val	-CH(CH$_3$)$_2$	X	-	-	-	X	-	Aliphatic	105
Tryptophan	W	Trp	-CH$_2$C$_8$H$_6$N	-	-	X	weak basic	-	-	Aromatic	163

Amino acid	Short	Abbrev.	Side chain	Hydro-phobic	pKa	Polar	pH	Small	Tiny	Aromatic or Aliphatic	van der Waals volume
Tyrosine	Y	Tyr	$-CH_2-C_6H_4OH$	-	10.46	X	weak acidic	-	-	Aromatic	141

Note: The pKa values of amino acids are typically slightly different when the amino acid is inside a protein. Protein pKa calculations are sometimes used to calculate the change in the pKa value of an amino acid in this situation.

Gene Expression and Biochemistry

Amino acid	Short	Abbrev.	Codon(s)	Occurrence in human proteins (%)	Essential‡ in humans
Alanine	A	Ala	GCU, GCC, GCA, GCG	7.8	No
Cysteine	C	Cys	UGU, UGC	1.9	Conditionally
Aspartic acid	D	Asp	GAU, GAC	5.3	No
Glutamic acid	E	Glu	GAA, GAG	6.3	Conditionally
Phenylalanine	F	Phe	UUU, UUC	3.9	Yes
Glycine	G	Gly	GGU, GGC, GGA, GGG	7.2	Conditionally
Histidine	H	His	CAU, CAC	2.3	Yes
Isoleucine	I	Ile	AUU, AUC, AUA	5.3	Yes
Lysine	K	Lys	AAA, AAG	5.9	Yes
Leucine	L	Leu	UUA, UUG, CUU, CUC, CUA, CUG	9.1	Yes
Methionine	M	Met	AUG	2.3	Yes
Asparagine	N	Asn	AAU, AAC	4.3	No
Pyrrolysine	O	Pyl	UAG*	0	No
Proline	P	Pro	CCU, CCC, CCA, CCG	5.2	No
Glutamine	Q	Gln	CAA, CAG	4.2	No
Arginine	R	Arg	CGU, CGC, CGA, CGG, AGA, AGG	5.1	Conditionally
Serine	S	Ser	UCU, UCC, UCA, UCG, AGU, AGC	6.8	No
Threonine	T	Thr	ACU, ACC, ACA, ACG	5.9	Yes
Selenocysteine	U	Sec	UGA**	>0	No
Valine	V	Val	GUU, GUC, GUA, GUG	6.6	Yes
Tryptophan	W	Trp	UGG	1.4	Yes
Tyrosine	Y	Tyr	UAU, UAC	3.2	Conditionally
Stop codon†	-	Term	UAA, UAG, UGA††	-	-

* UAG is normally the amber stop codon, but encodes pyrrolysine if a PYLIS element is present.

** UGA is normally the opal (or umber) stop codon, but encodes selenocysteine if a SECIS element is present.

† The stop codon is not an amino acid, but is included for completeness.

†† UAG and UGA do not always act as stop codons.

‡ An essential amino acid cannot be synthesized in humans and must, therefore, be supplied in the diet. Conditionally essential amino acids are not normally required in the diet, but must be supplied exogenously to specific populations that do not synthesize it in adequate amounts.

Mass Spectrometry

In mass spectrometry of peptides and proteins, knowledge of the masses of the residues is useful. The mass of the peptide or protein is the sum of the residue masses plus the mass of water.

Amino Acid	Short	Abbrev.	Formula	Mon. Mass§ (Da)	Avg. Mass (Da)
Alanine	A	Ala	C_3H_5NO	71.03711	71.0788
Cysteine	C	Cys	C_3H_5NOS	103.00919	103.1388
Aspartic acid	D	Asp	$C_4H_5NO_3$	115.02694	115.0886
Glutamic acid	E	Glu	$C_5H_7NO_3$	129.04259	129.1155
Phenylalanine	F	Phe	C_9H_9NO	147.06841	147.1766
Glycine	G	Gly	C_2H_3NO	57.02146	57.0519
Histidine	H	His	$C_6H_7N_3O$	137.05891	137.1411
Isoleucine	I	Ile	$C_6H_{11}NO$	113.08406	113.1594
Lysine	K	Lys	$C_6H_{12}N_2O$	128.09496	128.1741
Leucine	L	Leu	$C_6H_{11}NO$	113.08406	113.1594
Methionine	M	Met	C_5H_9NOS	131.04049	131.1986
Asparagine	N	Asn	$C_4H_6N_2O_2$	114.04293	114.1039
Pyrrolysine	O	Pyl	$C_{12}H_{21}N_3O_3$	255.15829	255.3172
Proline	P	Pro	C_5H_7NO	97.05276	97.1167
Glutamine	Q	Gln	$C_5H_8N_2O_2$	128.05858	128.1307
Arginine	R	Arg	$C_6H_{12}N_4O$	156.10111	156.1875
Serine	S	Ser	$C_3H_5NO_2$	87.03203	87.0782
Threonine	T	Thr	$C_4H_7NO_2$	101.04768	101.1051
Selenocysteine	U	Sec	C_3H_5NOSe	150.95364	150.0388
Valine	V	Val	C_5H_9NO	99.06841	99.1326
Tryptophan	W	Trp	$C_{11}H_{10}N_2O$	186.07931	186.2132
Tyrosine	Y	Tyr	$C_9H_9NO_2$	163.06333	163.1760

§ Monoisotopic mass

Stoichiometry and Metabolic Cost in Cell

The table below lists the abundance of amino acids in *E.coli* cells and the metabolic cost (ATP) for synthesis the amino acids. Negative numbers indicate the metabolic processes are energy favorable and do not cost net ATP of the cell. The abundance of amino acids includes amino acids in free form and in polymerization form (proteins).

Amino acid	Abundance (# of molecules ($\times 10^8$) per E. coli cell)	ATP cost in synthesis under aerobic condition	ATP cost in synthesis under anaerobic condition
Alanine	2.9	-1	1
Cysteine	0.52	11	15
Aspartic acid	1.4	0	2
Glutamic acid	1.5	-7	-1
Phenylalanine	1.1	-6	2
Glycine	3.5	-2	2
Histidine	0.54	1	7
Isoleucine	1.7	7	11
Lysine	2.0	5	9
Leucine	2.6	-9	1
Methionine	0.88	21	23
Asparagine	1.4	3	5
Proline	1.3	-2	4
Glutamine	1.5	-6	0
Arginine	1.7	5	13
Serine	1.2	-2	2
Threonine	1.5	6	8
Tryptophan	0.33	-7	7
Tyrosine	0.79	-8	2
Valine	2.4	-2	2

Remarks

Amino Acid	Abbrev.		Remarks
Alanine	A	Ala	Very abundant and very versatile, it is more stiff than glycine, but small enough to pose only small steric limits for the protein conformation. It behaves fairly neutrally, and can be located in both hydrophilic regions on the protein outside and the hydrophobic areas inside.
Asparagine or aspartic acid	B	Asx	A placeholder when either amino acid may occupy a position

Cysteine	C	Cys	The sulfur atom bonds readily to heavy metal ions. Under oxidizing conditions, two cysteines can join together in a disulfide bond to form the amino acid cystine. When cystines are part of a protein, insulin for example, the tertiary structure is stabilized, which makes the protein more resistant to denaturation; therefore, disulfide bonds are common in proteins that have to function in harsh environments including digestive enzymes (e.g., pepsin and chymotrypsin) and structural proteins (e.g., keratin). Disulfides are also found in peptides too small to hold a stable shape on their own (e.g. insulin).
Aspartic acid	D	Asp	Asp behaves similarly to glutamic acid, and carries a hydrophilic acidic group with strong negative charge. Usually, it is located on the outer surface of the protein, making it water-soluble. It binds to positively charged molecules and ions, and is often used in enzymes to fix the metal ion. When located inside of the protein, aspartate and glutamate are usually paired with arginine and lysine.
Glutamic acid	E	Glu	Glu behaves similarly to aspartic acid, and has a longer, slightly more flexible side chain.
Phenylalanine	F	Phe	Essential for humans, phenylalanine, tyrosine, and tryptophan contain a large, rigid aromatic group on the side chain. These are the biggest amino acids. Like isoleucine, leucine, and valine, these are hydrophobic and tend to orient towards the interior of the folded protein molecule. Phenylalanine can be converted into tyrosine.
Glycine	G	Gly	Because of the two hydrogen atoms at the α carbon, glycine is not optically active. It is the smallest amino acid, rotates easily, and adds flexibility to the protein chain. It is able to fit into the tightest spaces, e.g., the triple helix of collagen. As too much flexibility is usually not desired, as a structural component, it is less common than alanine.
Histidine	H	His	His is essential for humans. In even slightly acidic conditions, protonation of the nitrogen occurs, changing the properties of histidine and the polypeptide as a whole. It is used by many proteins as a regulatory mechanism, changing the conformation and behavior of the polypeptide in acidic regions such as the late endosome or lysosome, enforcing conformation change in enzymes. However, only a few histidines are needed for this, so it is comparatively scarce.
Isoleucine	I	Ile	Ile is essential for humans. Isoleucine, leucine, and valine have large aliphatic hydrophobic side chains. Their molecules are rigid, and their mutual hydrophobic interactions are important for the correct folding of proteins, as these chains tend to be located inside of the protein molecule.
Leucine or isoleucine	J	Xle	A placeholder when either amino acid may occupy a position

Lysine	K	Lys	Lys is essential for humans, and behaves similarly to arginine. It contains a long, flexible side chain with a positively charged end. The flexibility of the chain makes lysine and arginine suitable for binding to molecules with many negative charges on their surfaces. E.g., DNA-binding proteins have their active regions rich with arginine and lysine. The strong charge makes these two amino acids prone to be located on the outer hydrophilic surfaces of the proteins; when they are found inside, they are usually paired with a corresponding negatively charged amino acid, e.g., aspartate or glutamate.
Leucine	L	Leu	Leu is essential for humans, and behaves similarly to isoleucine and valine.
Methionine	M	Met	Met is essential for humans. Always the first amino acid to be incorporated into a protein, it is sometimes removed after translation. Like cysteine, it contains sulfur, but with a methyl group instead of hydrogen. This methyl group can be activated, and is used in many reactions where a new carbon atom is being added to another molecule.
Asparagine	N	Asn	Similar to aspartic acid, Asn contains an amide group where Asp has a carboxyl.
Pyrrolysine	O	Pyl	Similar to lysine, but it has a pyrroline ring attached.
Proline	P	Pro	Pro contains an unusual ring to the N-end amine group, which forces the CO-NH amide sequence into a fixed conformation. It can disrupt protein folding structures like α helix or β sheet, forcing the desired kink in the protein chain. Common in collagen, it often undergoes a post-translational modification to hydroxyproline.
Glutamine	Q	Gln	Similar to glutamic acid, Gln contains an amide group where Glu has a carboxyl. Used in proteins and as a storage for ammonia, it is the most abundant amino acid in the body.
Arginine	R	Arg	Functionally similar to lysine
Serine	S	Ser	Serine and threonine have a short group ended with a hydroxyl group. Its hydrogen is easy to remove, so serine and threonine often act as hydrogen donors in enzymes. Both are very hydrophilic, so the outer regions of soluble proteins tend to be rich with them.
Threonine	T	Thr	Essential for humans, Thr behaves similarly to serine.
Selenocysteine	U	Sec	The selenated form of cysteine, which replaces sulfur
Valine	V	Val	Essential for humans, Val behaves similarly to isoleucine and leucine.
Tryptophan	W	Trp	Essential for humans, Trp behaves similarly to phenylalanine and tyrosine. It is a precursor of serotonin and is naturally fluorescent.

Unknown	X	Xaa	Placeholder when the amino acid is unknown or unimportant
Tyrosine	Y	Tyr	Tyr behaves similarly to phenylalanine (precursor to tyrosine) and tryptophan, and is a precursor of melanin, epinephrine, and thyroid hormones. Naturally fluorescent, its fluorescence is usually quenched by energy transfer to tryptophans.
Glutamic acid or glutamine	Z	Glx	A placeholder when either amino acid may occupy a position

Catabolism

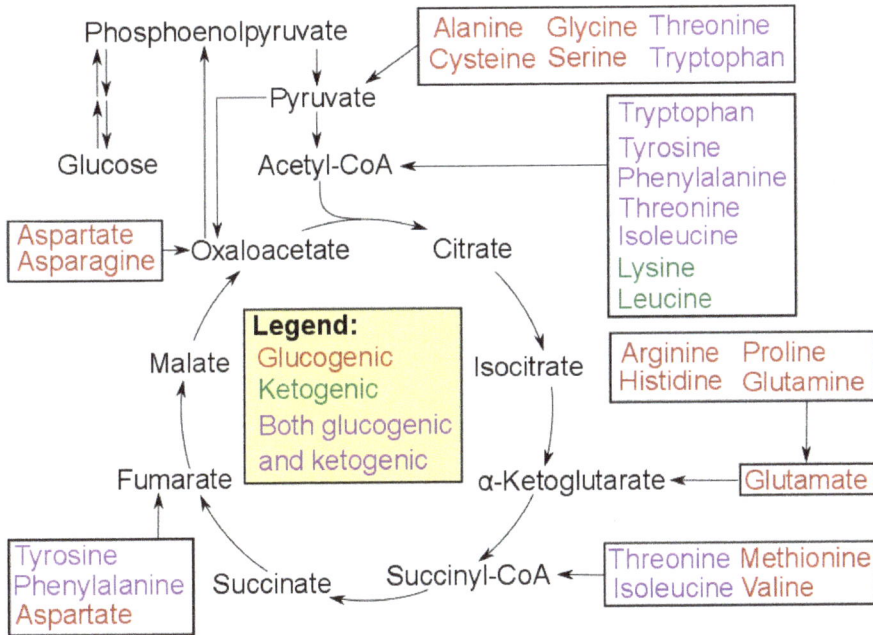

Amino acids can be classified according to the properties of their main products as either of:

- Glucogenic, with the products having the ability to form glucose by gluconeo-genesis
- Ketogenic, with the products not having the ability to form glucose: These prod-ucts may still be used for ketogenesis or lipid synthesis.
- Amino acids catabolized into both glucogenic and ketogenic products.

Life Based on Alternative Proteinogenic Sets

The proteinogenic set used by known life on Earth appears to be arbitrarily selected by evolution, according to current knowledge, from many hundreds of possible alpha-type amino acids. Xenobiology studies hypothetical life forms that could be constructed using alternative sets using expanded genetic codes. Miller-type experiments on artificial abio-genesis show that alpha-type amino acids predominate in water-based 'primordial soups', but beta-type amino acids dominate when less water is present. Both alpha- and beta-based sets could form the basis for alternative protein constructions and life forms.

Non-Proteinogenic Amino Acids

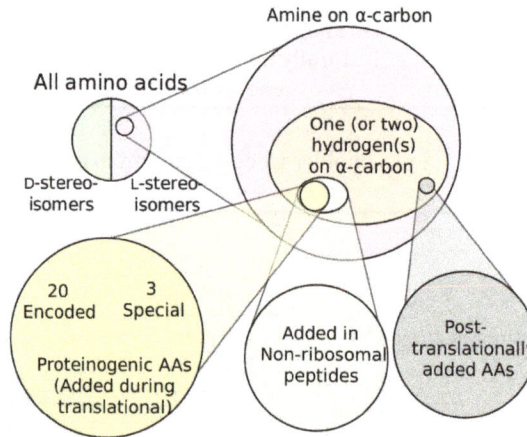

Proteinogenic amino acids are a small fraction of all amino acids

In biochemistry, non-coded, non-proteinogenic, or "unnatural" amino acids are those not naturally encoded or found in the genetic code of any organisms. Despite the use of only 23 amino acids (21 in eukaryotes) by the translational machinery to assemble proteins (the proteinogenic amino acids), over 140 natural amino acids are known and thousands of more combinations are possible. Several non-proteinogenic amino acids are noteworthy because they are:

- intermediates in biosynthesis

- post-translationally incorporated into protein

- possess a physiological role (e.g. components of bacterial cell walls, neurotransmitters and toxins)

- natural and man-made pharmacological compounds

- present in meteorites and in prebiotic experiments (e.g. Miller–Urey experiment)

Definition by Negation

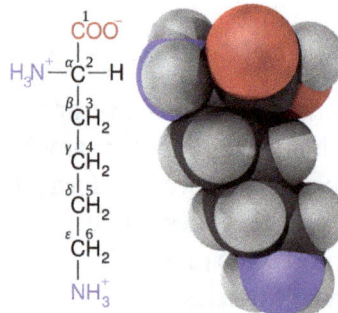

Lysine

Technically, any organic compound with an amine (-NH$_2$) and a carboxylic acid (-COOH) functional group is an amino acid. The proteinogenic amino acids are small subset of this group that possess central carbon atom (α- or 2-) bearing an amino group, a carboxyl group, a side chain and an α-hydrogen levo conformation, with the exception of glycine, which is achiral, and proline, whose amine group is a secondary amine and is consequently frequently referred to as an imino acid for traditional reasons, albeit not an imino.

The genetic code encodes 20 standard amino acids. However, there are three extra proteinogenic amino acids: selenocysteine, pyrrolysine and N-formylmethionine. The former two do not have a dedicated codon, but are added in place of a stop codon when a specific sequence is present, UGA codon and SECIS element for selenocysteine, UAG PYLIS downstream sequence for pyrrolysine. Formylmethionine is an amino acid encoded by the start codon AUG in bacteria, mitochondria and chloroplasts, but is often removed posttranslationally.

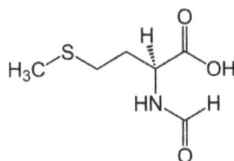

Formylmethionine. This amino acid is a methionine whose amino group has been protected by a formyl group

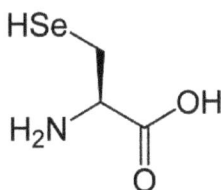

Selenocysteine. This amino acid contains a selenol group on its β-carbon

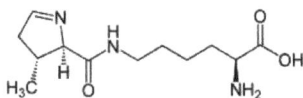

Pyrrolysine. This amino acid is formed by joining to the ε-amino group of lysine a carboxylated pyrroline ring

There are various groups of amino acids:

- 20 standard amino acids

- 23 proteinogenic amino acids

- over 80 amino acids created abiotically in high concentrations

- about 900 are produced by natural pathways

- over 118 engineered amino acids have been placed into protein

These groups overlap, but are not identical. All 23 proteinogenic amino acids are bio-synthesised by organisms, but not all of them are abiotic (found in prebiotic experiments and meteorites), such as histidine. Many amino acids, such as ornithine, are metabolic intermediates produced biotically, but not coded. Others are only metabolic intermediates, such as citrulline. Others are solely found in abiotic mixes, such as α-methylnorvaline. Over 30 unnatural amino acids have been translationally inserted into protein in engineered systems, yet are not biosynthetic.

Nomenclature

In addition to the IUPAC numbering system to differentiate the various carbons in an organic molecule, by sequentially assigning a number to each carbon, including those forming a carboxylic group, the carbons along the side-chain of amino acids can also be labelled with Greek letters, where the α-carbon is the central chiral carbon possessing a carboxyl group, a side chain and, in α-amino acids, an amino group – the carbon in carboxylic groups is not counted. (Consequently, the IUPAC names of many non-proteinogenic α-amino acids start with *2-amino-* and end in *-ic acid*.)

Natural, but Non L-α-amino Acids

Most natural amino acids are α-amino acids in the L conformation, but some exceptions exist.

Non-alpha

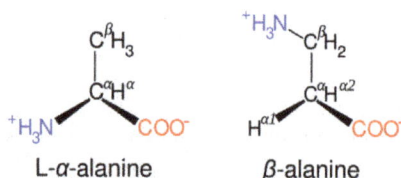

L-α-alanine   β-alanine

Some non-α amino acids exist in organisms. In these structures, the amine group displaced further from the carboxylic acid end of the amino acid molecule. Thus a β amino acid has the amine group bonded to the second carbon away, and a γ amino acid has it on the third. Examples include β-alanine, GABA, and δ-aminolevulinic acid.

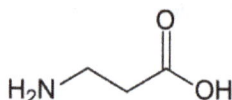

β-alanine: an amino acid produced by aspartate 1-decarboxylase and a precursor to coenzyme

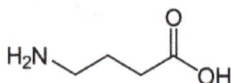

γ-Aminobutyric acid (GABA): a neurotransmitter in animals.

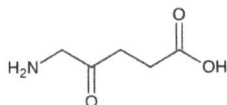

δ-Aminolevulinic acid: an intermediate in tetrapyrrole biosynthesis (haem, chlorophyll, cobalamin *etc.*).

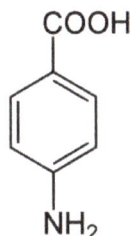

4-Aminobenzoic acid (PABA): an intermediate in folate biosynthesis

The reason why α-amino acids are used in proteins has been linked to their frequency in meteorites and prebiotic experiments. An initial speculation on the deleterious properties of β-amino acids in terms of secondary structure, turned out to be incorrect.

D-amino Acids

Some amino acids contain the opposite absolute chirality, chemicals that are not available from normal ribosomal translation/transcription machinery. Most bacterial cells walls are formed by peptidoglycan, a polymer composed of amino sugars crosslinked with short oligopeptides bridged between each other. The oligopeptide is non-ribosomally synthesised and contains several peculiarities, including D-amino acids, generally D-alanine and D-glutamate. A further peculiarity is that the former is racemised by a PLP-binding enzymes (encoded by *alr* or the homologue *dadX*), whereas the latter is racemised by a cofactor independent enzyme (*murI*). Some variants are present, in *Thermotoga* spp. D-lysine is present and in certain vancomycin-resistant bacteria D-serine is present (*vanT* gene).

In animals, some D-amino acids are neurotransmitters.

Without a Hydrogen on the α-carbon

All proteinogenic amino acids have at least one hydrogen on the α-carbon. Glycine has two hydrogens, and all others have one hydrogen and one side-chain. Replacement of the remaining hydrogen with a larger substituent, such as a methyl group, distorts the protein backbone.

In some fungi α-amino isobutyric acid is produced as a precursor to peptides, some of which exhibit antibiotic properties. This compound is similar to alanine, but possesses an additional methyl group on the α-carbon instead of a hydrogen. It is therefore achiral. Another compound similar to alanine without an α-hydrogen is dehydroalanine, which possess a methylene sidechain. It is one of several naturally occurring dehydroamino acids.

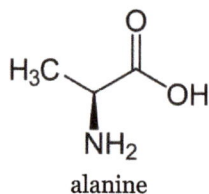

alanine

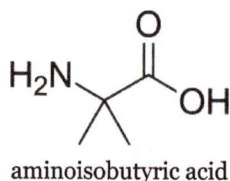

aminoisobutyric acid

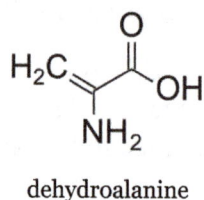

dehydroalanine

Twin Amino Acid Stereocentres

A subset of L-α-amino acids are ambiguous as to which of two ends is the α-carbon. In proteins a cysteine residue can form a disulfide bond with another cysteine residue, thus crosslinking the protein. Two crosslinked cysteines form a cystine molecule. Cysteine and methionine are generally produced by direct sulfurylation, but in some species they can be produced by transsulfuration, where the activated homoserine or serine is fused to a cysteine or homocysteine forming cystathionine. A similar compound is lanthionine, which can be seen as two alanine molecules joined via a thioether bond and is found in various organisms. Similarly, djenkolic acid, a plant toxin from jengkol beans, is composed of two cysteines connected by a methylene group. Diaminopimelic acid is both used as a bridge in petidoglycan and is used a precursor to lysine (via its decarboxylation).

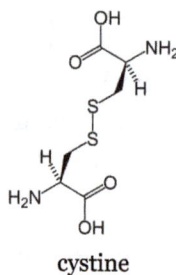

cystine

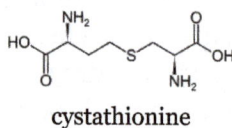

cystathionine

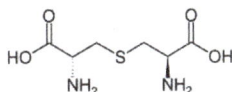
lanthionine

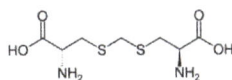
Djenkolic acid

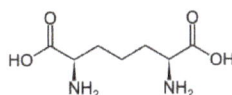
Diaminopimelic acid

Prebiotic Amino Acids and Alternative Biochemistries

In meteorites and in prebiotic experiments (e.g. Miller–Urey experiment) many more amino acids than the twenty standard amino acids are found, several of which at higher concentrations that the standard ones: it has been conjectured that if amino acid based life were to arise in parallel elsewhere in the universe, no more than 75% of the amino acids would be in common. The most notable anomaly is the lack of aminobutyric acid.

Proportion of amino acids relative to glycine (%)		
Molecule	Electric discharge	Murchinson meteorite
Glycine	100	100
Alanine	180	36
α-Amino-n-butyric acid	61	19
Norvaline	14	14
Valine	4.4	
Norleucine	1.4	
Leucine	2.6	
Isoleucine	1.1	
Alloisoleucine	1.2	
t-leucine	< 0.005	
α-Amino-n-heptanoic acid	0.3	
Proline	0.3	22
Pipecolic acid	0.01	11
α,β-diaminopropionic acid	1.5	
α,γ-diaminobutyric acid	7.6	
Ornithine	< 0.01	
lysine	< 0.01	

Proportion of amino acids relative to glycine (%)		
Molecule	Electric discharge	Murchinson meteorite
Aspartic acid	7.7	13
Glutamic acid	1.7	20
Serine	1.1	
Threonine	0.2	
Allothreonine	0.2	
Methionine	0.1	
Homocysteine	0.5	
Homoserine	0.5	
β-Alanine	4.3	10
β-Amino-n-butyric acid	0.1	5
β-Aminoisobutyric acid	0.5	7
γ-Aminobutyric acid	0.5	7
α-Aminoisobutyric acid	7	33
isovaline	1	11
Sarcosine	12.5	7
N-ethyl glycine	6.8	6
N-propyl glycine	0.5	
N-isopropyl glycine	0.5	
N-methyl alanine	3.4	3
N-ethyl alanine	< 0.05	
N-methyl β-alanine	1.0	
N-ethyl β-alanine	< 0.05	
isoserine	1.2	
α-hydroxy-γ-aminobutyric acid	17	

Straight Side Chain

The genetic code has been described as a frozen accident and the reasons why there is only one standard amino acid with a straight chain (alanine) could simply be redundancy with valine, leucine and isoleucine. However, straight chained amino acids are reported to form much more stable alpha helices.

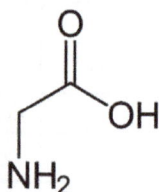

Glycine (Hydrogen side-chain)

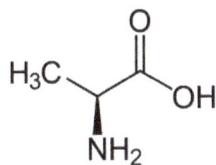
Alanine (Methyl side-chain)

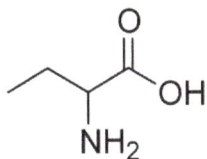
α-aminobutyric acid (Ethyl side-chain)

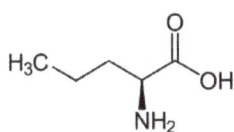
Norvaline (*n*-Propyl side-chain)

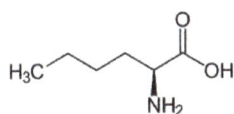
Norleucine (*n*-Butyl side-chain)

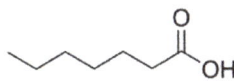
Homonorleucine (*n*-Pentyl side-chain)

Chalcogen

Serine, homoserine, O-methyl-homoserine and O-ethyl-homoserine possess an hydroxymethyl, hydroxyethyl, O-methyl-hydroxymethyl and O-methyl-hydroxyethyl side chain. Whereas cysteine, homocysteine, methionine and ethionine possess the thiol equivalents. The selenol equivalents are selenocysteine, selenohomocysteine, selenomethionine and selenoethionine. Amino acids with the next chalcogen down are also found in nature: several species such as Aspergillus fumigatus, Aspergillus terreus, and Penicillium chrysogenum in the absence of sulfur are able to produce and incorporate into protein tellurocysteine and telluromethionine.

Hydroxyglycine, an amino acid with a hydroxyl side-chain, is highly unstable.

Expanded Genetic Code

Roles

In cells, especially autotrophs, several non-proteinogenic amino acids are found as metabolic intermediates. However, despite the catalytic flexibility of PLP-binding enzymes, many amino acids are synthesised as keto-acids (*e.g.* 4-methyl-2-oxopentanoate to leucine) and aminated in the last step, thus keeping the number of non-proteinogenic amino acid intermediates fairly low.

Ornithine and citrulline occur in the urea cycle, part of amino acid catabolism.

In addition to primary metabolism, several non-proteinogenic amino acids are precursors or the final production in secondary metabolism to make small compounds or non-ribosomal peptides (such as some toxins).

Post-translationally Incorporated into Protein

Despite not being encoded by the genetic code as proteinogenic amino acids, some non-standard amino acids are nevertheless found in proteins. These are formed by post-translational modification of the side chains of standard amino acids present in the target protein. These modifications are often essential for the function or regulation of a protein; for example, in Gamma-carboxyglutamate the carboxylation of glutamate allows for better binding of calcium cations, and in hydroxyproline the hydroxylation of proline is critical for maintaining connective tissues. Another example is the formation of hypusine in the translation initiation factor EIF5A, through modification of a lysine residue. Such modifications can also determine the localization of the protein, e.g., the addition of long hydrophobic groups can cause a protein to bind to a phospholipid membrane.

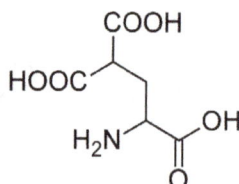

Carboxyglutamic acid. Whereas glutamic acid possess one γ-carboxyl group, Carboxyglutamic acid possess two.

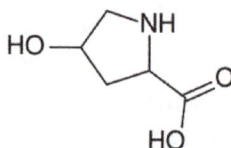

Hydroxyproline. This imino acid differs from proline due to a hydroxyl group on carbon 4.

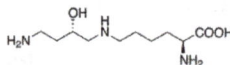

Hypusine. This amino acid is obtained by adding to the ε-amino group of a lysine a 4-aminobutyl moiety (obtained from spermidine)

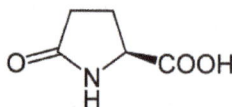

Pyroglutamic acid

There is some preliminary evidence that aminomalonic acid may be present, possibly by misincorporation, in protein.

Toxic Analogues

Several non-proteinogenic amino acids are toxic due to their ability to mimic certain properties of proteinogenic amino acids, such as thialysine. Some non-proteinogenic amino acids are neurotoxic by mimicking amino acids used as neurotransmitters (i.e. not for protein biosynthesis), e.g. Quisqualic acid, canavanine or azetidine-2-carboxylic acid. Cephalosporin C has an α-aminoadipic acid (homoglutamate) backbone that is amidated with a cephalosporin moiety. Penicillamine is therapeutic amino acid, whose mode of action is unknown.

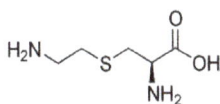
Thialysine

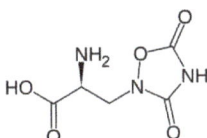
Quisqualic acid

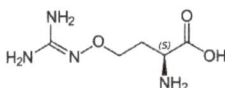
Canavanine

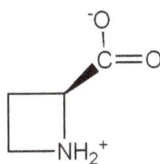
azetidine-2-carboxylic acid

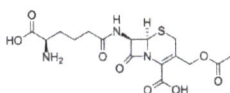
Cephalosporin C

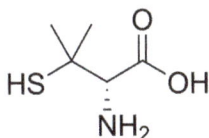
Penicillamine

Naturally-occurring cyanotoxins can also include non-proteinogenic amino acids. Microcystin and nodularin, for example, are both derived from ADDA, a β-amino acid.

Not Amino Acids

Taurine is an amino sulfonic acid and not an amino acid, however it is occasionally considered as such as the amounts required to suppress the auxotroph in certain organisms (e.g. cats) are closer to those of "essential amino acids" (amino acid auxotrophy) than of vitamins (cofactor auxotrophy).

The osmolytes, sarcosine and glycine betaine are derived from amino acids, but have a secondary and quaternary amine respectively.

Beta-Alanine

β-Alanine (or *beta*-alanine) is a naturally occurring beta amino acid, which is an amino acid in which the amino group is at the β-position from the carboxylate group (i.e., two atoms away, see Figure 1). The IUPAC name for β-alanine is 3-aminopropanoic acid. Unlike its counterpart α-alanine, β-alanine has no stereocenter.

β-Alanine is not used in the biosynthesis of any major proteins or enzymes. It is formed in vivo by the degradation of dihydrouracil and carnosine. It is a component of the naturally occurring peptides carnosine and anserine and also of pantothenic acid (vitamin B_5), which itself is a component of coenzyme A. Under normal conditions, β-alanine is metabolized into acetic acid.

β-Alanine is the rate-limiting precursor of carnosine, which is to say carnosine levels are limited by the amount of available β-alanine, not histidine. Supplementation with β-alanine has been shown to increase the concentration of carnosine in muscles, decrease fatigue in athletes and increase total muscular work done. Simply supplementing with carnosine is not as effective as supplementing with β-alanine alone since carnosine, when taken orally, is broken down during digestion to its components, histidine and β-alanine. Hence, by weight, only about 40% of the dose is available as β-alanine.

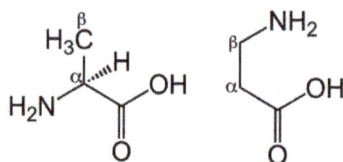

Figure 1: Comparison of β-alanine (right) with the more customary (chiral) amino acid, L-α-alanine (left)

L-Histidine, with a pK_a of 6.1 is a relatively weak buffer over the physiological intramuscular pH range. However, when bound to other amino acids, this increases nearer to 6.8-7.0. In particular, when bound to β-alanine, the pK_a value is 6.83, making this a very efficient intramuscular buffer. Furthermore, because of the position of the beta amino group, β-alanine dipeptides are not incorporated into proteins, and thus can be stored at relatively high concentrations (millimolar). Occurring at 17−25 mmol/kg (dry muscle), carnosine (β-alanyl-L-histidine) is an important intramuscular buffer, constituting 10-20% of the total buffering capacity in type I and II muscle fibres.

Even though much weaker than glycine (and, thus, with a debated role as a physiological transmitter), β-alanine is an agonist next in activity to the cognate ligand glycine itself, for strychnine-sensitive inhibitory glycine receptors (GlyRs) (the agonist order: glycine ≫ β-alanine > taurine ≫ alanine, L-serine > proline).

Athletic Performance Enhancement

Daily β-alanine supplementation can safely increase exercise performance, especially in exercise tasks lasting 1–4 minutes.

Ingestion of β-Alanine can cause paraesthesia, reported as a tingling sensation, in a dose-dependent fashion.

Other uses

A high-potency artificial sweetener, called suosan, is derived from β-alanine.

Amino Acid Neurotransmitter

An amino acid neurotransmitter is an amino acid which is able to transmit a nerve message across a synapse. Neurotransmitters (chemicals) are packaged into vesicles that cluster beneath the axon terminal membrane on the presynaptic side of a synapse in a process called endocytosis.

Amino acid neurotransmitter release (exocytosis) is dependent upon calcium Ca^{2+} and is a presynaptic response. There are inhibitory amino acids (IAA) or excitatory amino acids (EAA). Some EAA are L-Glutamate, L-Aspartate, L-Cysteine, and L-Homocysteine. These neurotransmitter systems will activate post-synaptic cells. Some IAA include GABA, Glycine, β-Alanine, and Taurine. The IAA depress the activity of post-synaptic cells.

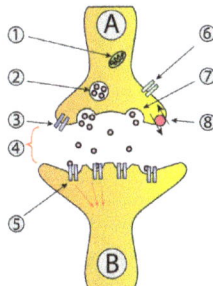

Activity at an axon terminal: Neuron A is transmitting a signal at the axon terminal to neuron B (receiving). Features: 1. Mitochondrion. 2. synaptic vesicle with neurotransmitters. 3. Autoreceptor. 4. Synapse with neurotransmitter released (serotonin). 5. Postsynaptic receptors activated by neurotransmitter (induction of a postsynaptic potential). 6. Calcium channel. 7. Exocytosis of a vesicle. 8. Recaptured neurotransmitter.

Aminocaproic Acid

Aminocaproic acid (also known as ε-aminocaproic acid, ε-Ahx, or 6-aminohexanoic acid) is a derivative and analogue of the amino acid lysine, which makes it an effective inhibitor for enzymes that bind that particular residue. Such enzymes include proteolytic enzymes like plasmin, the enzyme responsible for fibrinolysis. For this reason it is effective in treatment of certain bleeding disorders, and it is marketed as Amicar. Aminocaproic acid is also an intermediate in the polymerization of Nylon-6, where it is formed by ring-opening hydrolysis of caprolactam.

Clinical use

Aminocaproic acid is used to treat excessive postoperative bleeding, especially after procedures in which a great amount of bleeding is expected, such as cardiac surgery. It can be given orally or intravenously. A meta-analysis found that lysine analogs like aminocaproic acid significantly reduced blood loss in patients undergoing coronary artery bypass grafting.

Aminocaproic acid can be used to treat the overdose and/or toxic effects of the thrombolytic pharmacologic agents tissue plasminogen activator (commonly known as tPA) and streptokinase.

Bleeding Due to Elevated Fibrinolytic Activity

Aminocaproic acid is used for the treatment of excessive bleeding resulting from systemic hyperfibrinolysis and urinary fibrinolysis. In life-threatening situations, fresh whole blood, fibrinogen infusions, and other emergency measures also may be required.

Aminocaproic acid is used in systemic hyperfibrinolysis associated with surgical complications following heart surgery (with or without cardiac bypass procedures) and portacaval shunt; in carcinoma of the lung, prostate, cervix, or stomach; in abruptio placentae; and in hematologic disorders such as amegakaryocytic thrombocytopenia accompanying aplastic anemia (reduces the need for platelet transfusions).

It is used in urinary fibrinolysis associated with complications of severe trauma, anoxia, and shock, and as manifested by surgical hematuria especially following prostatectomy and nephrectomy, or in nonsurgical hematuria accompanying polycystic or neoplastic disease of the GU tract.

It is used in conjunction with heparin therapy in patients with acute promyelocytic leukemia. Use is not currently included in the labeling approved by the US Food and Drug Administration. It is used as initiate therapy when plasma α2-antiplasmin (α2-plasmin inhibitor) levels have decreased to <40% of normal levels.

Ocular Hemorrhage

According to the Cochrane database, aminocaproic acid has been used effectively for the prevention of secondary ocular hemorrhage in patients with nonperforating traumatic hyphema. (Use is not currently included in the labeling approved by the US Food and Drug Administration.)

The US FDA has designated it an orphan drug for topical treatment of traumatic hyphema.

Hereditary Hemorrhagic Telangiectasia

Aminocaproic acid has been used orally for the management of hereditary hemorrhagic telangiectasia. This use is not currently included in the labeling approved by the US Food and Drug Administration.

Side Effects

Side effects include Confusion, Vision decrease, Vomiting, Headache, Convulsions, Malaise, Muscle weakness, Dizziness, Tinnitus, Nausea, Bradycardia, Thrombosis, Edema, Hypotension, Stroke, Syncope, Intracranial hypertension, Peripheral ischemia, Pulmonary embolism, Dyspnea, Congestion, Diarrhea, Abdominal pain, Leukopenia, Agranulocytosis, Coagulation disorder, Dry ejaculation, Injection site reactions (pain/ necrosis), Pruritus, Rash, Renal failure, Anaphylaxis, and risk for myalgia.

Alpha-Aminobutyric Acid

α-Aminobutyric acid (AABA), also known as homoalanine in biochemistry, is a non-proteinogenic alpha amino acid with chemical formula $C_4H_9NO_2$. The straight two carbon side chain is one carbon longer than alanine, hence the prefix homo-.

Homoalaine is biosynthesised by transaminating oxobutyrate, a metabolite in isoleucine biosynthesis. It is used by nonribosomal peptide synthases. One example of a nonribosomal peptide containing homoalanine is ophthalmic acid, which was first isolated from calf lens.

α-Aminobutyric acid is an isomer of the amino acid aminobutyric acid. It has two other isomers, the neurotransmitter γ-Aminobutyric acid (GABA) and β-Aminobutyric acid (BABA) which is known for inducing plant disease resistance.

Gamma-Aminobutyric Acid

gamma-Aminobutyric acid is the chief inhibitory neurotransmitter in the mammalian central nervous system. It plays the principal role in reducing neuronal excitability throughout the nervous system. In humans, GABA is also directly responsible for the regulation of muscle tone.

Although in chemical terms it is an amino acid, GABA is rarely referred to as such in the scientific or medical communities, because the term "amino acid," used without a qualifier, by convention refers to the alpha amino acids, which GABA is not, nor is it considered to be incorporated into proteins.

Function

Neurotransmitter

In vertebrates, GABA acts at inhibitory synapses in the brain by binding to specific transmembrane receptors in the plasma membrane of both pre- and postsynaptic neuronal processes. This binding causes the opening of ion channels to allow the flow of either negatively charged chloride ions into the cell or positively charged potassium ions out of the cell. This action results in a negative change in the transmembrane potential, usually causing hyperpolarization. Two general classes of GABA receptor are known: $GABA_A$ in which the receptor is part of a ligand-gated ion channel complex, and $GABA_B$ metabotropic receptors, which are G protein-coupled receptors that open or close ion channels via intermediaries (G proteins).

The production, release, action, and degradation of GABA at a stereotyped GABAergic synapse

Neurons that produce GABA as their output are called GABAergic neurons, and have chiefly inhibitory action at receptors in the adult vertebrate. Medium Spiny Cells are a typical example of inhibitory CNS GABAergic cells. In contrast, GABA exhibits both excitatory and inhibitory actions in insects, mediating muscle activation at synapses between nerves and muscle cells, and also the stimulation of certain glands. In mammals, some GABAergic neurons, such as chandelier cells, are also able to excite their glutamatergic counterparts.

$GABA_A$ receptors are ligand-activated chloride channels; when activated by GABA, they allow the flow of chloride ions across the membrane of the cell. Whether this chloride flow is excitatory/depolarizing (makes the voltage across the cell's membrane less negative), shunting (has no effect on the cell's membrane potential) or inhibitory/hyperpolarizing (makes the cell's membrane more negative) depends on the direction of the flow of chloride. When net chloride flows out of the cell, GABA is excitatory or depolarizing; when chloride flows into the cell, GABA is inhibitory or hyperpolarizing. When the net flow of chloride is close to zero, the action of GABA is shunting. Shunting inhibition has no direct effect on the membrane potential of the cell; however, it reduces the effect of any coincident synaptic input by reducing the electrical resistance of the cell's membrane. A developmental switch in the molecular machinery controlling concentration of chloride inside the cell changes the functional role of GABA between neonatal and adult stages. As the brain develops into adulthood GABA's role changes from excitatory to inhibitory .

Brain Development

While GABA is an inhibitory transmitter in the mature brain, its actions are primarily excitatory in the developing brain. The gradient of chloride is reversed in immature neurons, and its reversal potential is higher than the resting membrane potential of the cell; activation of a GABA-A receptor thus leads to efflux of Cl^- ions from the cell, i.e. a depolarizing current. The differential gradient of chloride in immature neurons is primarily due to the higher concentration of NKCC1 co-transporters relative to KCC2 co-transporters in immature cells. GABA itself is partially responsible for orchestrating the maturation of ion pumps. GABA-ergic interneurons mature faster in the hippocampus and the GABA signalling machinery appears earlier than glutamatergic transmission. Thus, GABA is the major excitatory neurotransmitter in many regions of the brain before the maturation of glutamatergic synapses.

However, this theory has been questioned based on results showing that in brain slices of immature mice incubated in artificial cerebrospinal fluid (ACSF) (modified in a way that takes into account the normal composition of the neuronal milieu in sucklings by adding an energy substrate alternative to glucose, beta-hydroxybutyrate) GABA action shifts from excitatory to inhibitory mode.

This effect has been later repeated when other energy substrates, pyruvate and lactate, supplemented glucose in the slices' media. Later investigations of pyruvate and lactate metabolism found that the original results were not due to energy source issues but to changes in pH resulting from the substrates acting as "weak acids". These arguments were later rebutted by further findings showing that changes in pH even greater than that caused by energy substrates do not affect the GABA-shift described in the presence of energy substrate-fortified ACSF and that the mode of action of beta-hydroxybutyrate, pyruvate and lactate (assessed by measurement NAD(P)H and oxygen utilization) was energy metabolism-related.

In the developmental stages preceding the formation of synaptic contacts, GABA is synthesized by neurons and acts both as an autocrine (acting on the same cell) and paracrine (acting on nearby cells) signalling mediator. The ganglionic eminences also contribute greatly to building up the GABAergic cortical cell population.

GABA regulates the proliferation of neural progenitor cells the migration and differentiation the elongation of neurites and the formation of synapses.

GABA also regulates the growth of embryonic and neural stem cells. GABA can influence the development of neural progenitor cells via brain-derived neurotrophic factor (BDNF) expression. GABA activates the GABA$_A$ receptor, causing cell cycle arrest in the S-phase, limiting growth.

Beyond the Nervous System

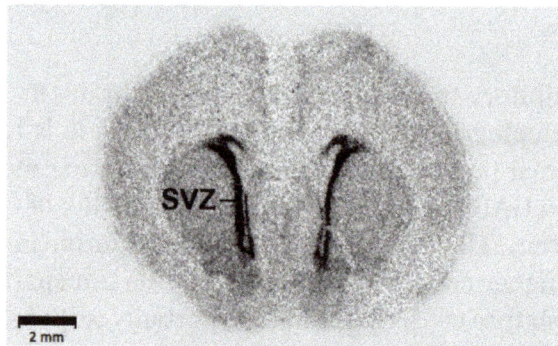

mRNA expression of the embryonic variant of the GABA-producing enzyme GAD67 in a coronal brain section of a one-day-old Wistar rat, with the highest expression in subventricular zone (svz); from Popp et al., 2009

GABAergic mechanisms have been demonstrated in various peripheral tissues and organs including, but not restricted to, the intestine, stomach, pancreas, Fallopian tube, uterus, ovary, testis, kidney, urinary bladder, lung, and liver.

In 2007, an excitatory GABAergic system was described in the airway epithelium. The system activates following exposure to allergens and may participate in the mechanisms of asthma. GABAergic systems have also been found in the testis and in the eye lens.

Structure and Conformation

GABA is found mostly as a zwitterion, that is, with the carboxy group deprotonated and the amino group protonated. Its conformation depends on its environment. In the gas phase, a highly folded conformation is strongly favored because of the electrostatic attraction between the two functional groups. The stabilization is about 50 kcal/mol, according to quantum chemistry calculations. In the solid state, a more extended conformation is found, with a trans conformation at the amino end and a gauche conformation at the carboxyl end. This is due to the packing interactions with the neighboring molecules. In solution, five different conformations, some folded and some extended, are found as a result of solvation effects. The conformational flexibility of GABA is important for its biological function, as it has been found to bind to different receptors with different conformations. Many GABA analogues with pharmaceutical applications have more rigid structures in order to control the binding better.

History

Gamma-aminobutyric acid was first synthesized in 1883, and was first known only as a plant and microbe metabolic product. In 1950, however, GABA was discovered to be an integral part of the mammalian central nervous system.

Biosynthesis

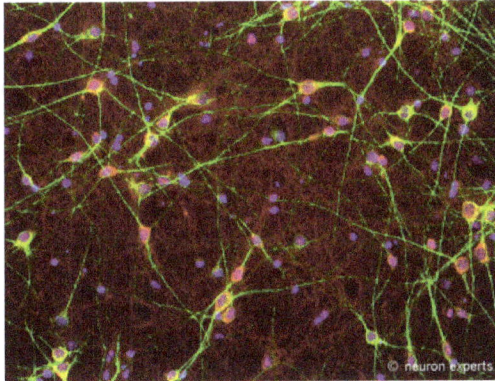

Gabaergic neurons which produce GABA

Exogenous GABA does not penetrate the blood–brain barrier; it is synthesized in the brain. It is synthesized from glutamate using the enzyme Glutamate decarboxylase (GAD) and pyridoxal phosphate (which is the active form of vitamin B6) as a cofactor. This process converts glutamate, the principal excitatory neurotransmitter, into the principal inhibitory neurotransmitter (GABA). GABA is converted back to glutamate by a metabolic pathway called the GABA shunt.

Catabolism

GABA transaminase enzyme catalyzes the conversion of 4-aminobutanoic acid (GABA) and 2-oxoglutarate (α-ketoglutarate) into succinic semialdehyde and glutamate. Succinic semialdehyde is then oxidized into succinic acid by succinic semialdehyde dehydrogenase and as such enters the citric acid cycle as a usable source of energy.

Pharmacology

Drugs that act as allosteric modulators of GABA receptors (known as GABA analogues or *GABAergic* drugs) or increase the available amount of GABA typically have relaxing, anti-anxiety, and anti-convulsive effects. Many of the substances below are known to cause anterograde amnesia and retrograde amnesia.

In general, GABA does not cross the blood–brain barrier, although certain areas of the brain that have no effective blood–brain barrier, such as the periventricular nucleus, can be reached by drugs such as systemically injected GABA. At least one study suggests that orally administered GABA increases the amount of Human Growth Hormone (HGH). GABA directly injected to the brain has been reported to have both stimulatory and inhibitory effects on the production of growth hormone, depending on the physiology of the individual. Certain pro-drugs of GABA (ex. picamilon) have been developed to permeate the blood–brain barrier, then separate into GABA and the carrier molecule once inside the brain. This allows for a direct increase of GABA levels throughout all areas of the brain, in a manner following the distribution pattern of the pro-drug prior to metabolism.

GABA enhanced the catabolism of serotonin into N-Acetylserotonin (the precursor of melatonin) in rats. It is thus suspected that GABA is involved in the synthesis of melatonin and thus may exert regulatory effects on sleep and reproductive functions.

GABAergic Drugs

- GABA$_A$ receptor ligands

    - Agonists/Positive allosteric modulators: ethanol, barbiturates, benzodiazepines, carisoprodol, chloral hydrate, etaqualone, etomidate, glutethimide, kava, methaqualone, muscimol, neuroactive steroids, z-drugs, propofol, skullcap, valerian, theanine, volatile/inhaled anaesthetics.

    - Antagonists/Negative allosteric modulators: bicuculline, cicutoxin, flumazenil, furosemide, gabazine, oenanthotoxin, picrotoxin, Ro15-4513, thujone, amentoflavone.

- GABA$_B$ receptor ligands

    - Agonists: baclofen, GBL, propofol, GHB, phenibut.

    - Antagonists: phaclofen, saclofen.

- GABA reuptake inhibitors: deramciclane, hyperforin, tiagabine.

- GABA-transaminase inhibitors: gabaculine, phenelzine, valproate, vigabatrin, lemon balm (*Melissa officinalis*).

- GABA analogues: pregabalin, 4-Methylpregabalin, gabapentin, gabapentin enacarbil, atagabalin, imagabalin, mirogabalin.

- Others: GABA (itself), L-glutamine, L-theanine, picamilon, progabide.

In Plants

GABA is also found in plants. It is the most abundant amino acid in the apoplast of tomatoes. It has also a role in cell signalling in plants.

Carnitine

Carnitine is an amino acid derivative and nutrient involved in lipid (fat) metabolism in mammals and other eukaryotes. It is in the chemical compound classes of β-hydroxyacids and quaternary ammonium compounds, and because of the hydroxyl-substituent, it exists in two stereoisomers, the biologically active enantiomer L-carnitine, and the essentially biologically inactive D-carnitine. Both are available through chemical synthesis, and the L-form is continuously biosynthesized in eukaryotic organisms from the proteinogenic amino acids lysine and methionine. In such eukaryotic cells, it is specifically required for the transport of fatty acids from the intermembraneous space in the mitochondria into the mitochondrial matrix during the catabolism of lipids, in the generation of metabolic energy. Carnitine was originally found as a growth factor for mealworms and labeled vitamin B_T, although carnitine is not by biochemical definition a true vitamin. It is used efficaciously, clinically, in the treatment of some conditions, e.g. systemic primary carnitine deficiency, and it is available over the counter as a nutritional supplement, though its efficacy for most conditions for which it is advertised is controversial or not yet established.

Biosynthesis and Metabolism

In animals, the biosynthesis of carnitine occurs primarily in the liver and kidneys from the amino acids lysine (via trimethyllysine) and methionine.

Carnitine transports long-chain acyl groups from fatty acids into the mitochondrial matrix, so they can be broken down through β-oxidation to acetyl CoA to obtain usable energy via the citric acid cycle. In some organisms, such as fungi, the acetate is used in the glyoxylate cycle for gluconeogenesis and formation of carbohydrates. Fatty acids must be activated before being covalently linked to the carnitine molecule to form acyl-carnitine for transport. The free fatty acid in the cytosol is first adenylated by reaction with ATP, then attached with a thioester bond to coenzyme A (CoA), with expulsion of AMP. This reaction is catalyzed by the enzyme fatty acyl-CoA synthetase and driven to completion by inorganic pyrophosphatase.

The acyl group on CoA can now be transferred to carnitine and the resulting acyl-carnitine transported into the mitochondrial matrix. This occurs via a series of similar steps:

- Acyl CoA is transferred to the hydroxyl group of carnitine by carnitine acyltransferase I (palmitoyltransferase) located on the outer mitochondrial membrane

- Acylcarnitine is shuttled inside by a carnitine-acylcarnitine translocase

- Acylcarnitine is converted to acyl CoA by carnitine acyltransferase II (palmitoyltransferase) located on the inner mitochondrial membrane. The liberated carnitine returns to the cytosol.

Carnitine acyltransferase I and peroxisomal carnitine octanoyl transferase (CROT) undergo allosteric inhibition by malonyl-CoA, an intermediate in fatty acid biosynthesis, to prevent futile cycling between β-oxidation and fatty acid synthesis.

Human genetic disorders, such as primary carnitine deficiency, carnitine palmitoyltransferase I deficiency, carnitine palmitoyltransferase II deficiency and carnitine-acyl-carnitine translocase deficiency, affect different steps of this process.

Atherosclerosis

A link between dietary consumption of carnitine and atherosclerosis has been proposed. There is evidence that carnitine lowers the risk of mortality due to arrythmias, after an acute myocardial infarction. When intestinal bacteria are exposed to carnitine from food, they produce the byproduct, trimethylamine, which is oxidised in the liver to trimethylamine N-oxide (TMAO), which may be associated with atherosclerosis; the risk of cardiovascular events is higher in those with high TMAO levels, independent of the observed level of carnitine. The presence of TMAO-producing bacteria is reportedly a consequence of subject's consuming diets rich in meat. Vegetarian and vegans who ate a single meal of meat had much lower levels of TMAO in their bloodstream than did regular meat-eaters, which was ascribed to the lower levels of the intestinal bacteria that convert carnitine into TMAO.

A further study reported evidence of a second path for atherogenic activity of carnitine, passing through a different metabolite, γ-butyrobetaine (γBB).

Effects on Bone Mass

Carnitine concentration in cells diminishes as humans age, affecting fatty acid metabolism in various tissues. Bone is affected, in particular, as it requires the continuous reconstructive and metabolic functions of osteoblasts for maintenance of its mass. A 2008 study reported that supplementing rat diets with L-carnitine decreased bone turnover and increased bone mineral density in ovariectomized females.

Effect on Thyroid Hormone Action

A 2001 report suggested that L-carnitine may be useful in preventing and reversing hyperthyroid symptoms. A 2004 study reported that L-carnitine acts as a peripheral antagonist of thyroid hormone action; in particular, L-carnitine inhibited both triiodo-thyronine (T_3) and thyroxine (T_4) entry into the cell nuclei.

Possible Health Effects

Carnitine has been proposed as a supplement to treat a variety of health conditions including heart attack, heart failure, angina, narcolepsy, and diabetic neuropathy, but not improving exercise performance, nor wasting syndrome (weight loss). In all of these cases the results are preliminary or proposed, and not part of an established medical treatment.

People with epileptic disorders/vulnerability are advised, on product monographs and pharmacy databases, to avoid carnitine and its derivatives as it is claimed to promote epileptic discharges. It now appears that this advice has been misguided. According to a systematic review of epileptics treated with valproic acid, these claims are unsubstantiated both in the scientific literature and in clinical practice and may have done more harm than good. This is unfortunate as many epileptics actually have decreased levels of carnitine (especially those under valproate treatment) and might need to supplement with this substance in order to avoid troubling side effects (e.g., hepatotoxicity, hyperammonemic encephalopathy). The exact reason for these cautionary notes against carnitine is not known. One possibility is that a well-intended, but ultimately false, disclaimer has propagated uncontrollably throughout pharmacy texts and electronic databases worldwide. Another possibility is that some of the side effects of carnitine (e.g., overstimulation, dizziness, headache, nausea) may mimic, and thus be mistaken for, a mild epileptic fit.

Sources

Food

The highest concentrations of carnitine are found in red meat. It can be found at significantly lower levels in many other foods including nuts and seeds (e.g. pumpkin, sunflower, sesame), legumes or pulses (beans, peas, lentils, peanuts), vegetables (artichokes, asparagus, beet greens (young leaves of the beetroot), broccoli, brussels sprouts, collard greens, garlic, mustard greens, okra, parsley, kale), fruits (apricots, bananas), cereals (buckwheat, corn, millet, oatmeal, rice bran, rye, whole wheat, wheat bran, wheat germ) and other foods (bee pollen, brewer's yeast, carob).

Product	Quantity	Carnitine
Lamb	100 g	190 mg
Beef steak	100 g	95 mg
Ground beef	100 g	94 mg

Pork	100 g	27.7 mg
Bacon	100 g	23.3 mg
Tempeh	100 g	19.5 mg
Cod fish	100 g	5.6 mg
Chicken breast	100 g	3.9 mg
American cheese	100 g	3.7 mg
Ice cream	100 mL	3.7 mg
Whole milk	100 mL	3.3 mg
Avocado	one medium	2 mg
Cottage cheese	100 g	1.1 mg
Whole-wheat bread	100 g	0.36 mg
Asparagus	100 g	0.195 mg
White bread	100 g	0.147 mg
Macaroni	100 g	0.126 mg
Peanut butter	100 g	0.083 mg
Rice (cooked)	100 g	0.0449 mg
Egg	100 g	0.0121 mg
Orange juice	100 mL	0.0019 mg
Lentil	100 g	2.1 mg
Potato	100 g	2.4 mg
Sweet Potato	100 g	1.1 mg
Banana	100 g	0.2 mg
Carrot	100 g	0.3 mg
Apple (without skin)	100 g	0.2 mg
Raisin	100 g	0.8 mg

In general, 20 to 200 mg are ingested per day by those on an omnivorous diet, whereas those on a strict vegetarian or vegan diet may ingest as little as 1 mg/day. However, even strict vegetarians (vegans) show no signs of carnitine deficiency, despite the fact that most dietary carnitine is derived from animal sources. No advantage appears to exist in giving an oral dose greater than 2 g at one time, since absorption studies indicate saturation at this dose.

Health Canada

Other sources may be found in over-the-counter vitamins, energy drinks and various other products. Products containing L-carnitine can now be marketed as "natural health products" in Canada. As of 2012, Parliament has allowed carnitine products and supplements to be imported into Canada (Health Canada). The Canadian government did issue an amendment in December 2011 allowing the sale of L-carnitine without a prescription.

History

Levocarnitine was approved by the U.S. Food and Drug Administration as a new molecular entity under the brand name Carnitor on December 27, 1985.

Citrulline

The organic compound citrulline is an α-amino acid. Its name is derived from citrullus, the Latin word for watermelon, from which it was first isolated in 1914 by Koga & Odake. It was finally identified by Wada in 1930. It has the formula $H_2NC(O)NH(CH_2)_3CH(NH_2)CO_2H$. It is a key intermediate in the urea cycle, the pathway by which mammals excrete ammonia.

In the body, citrulline is produced as a byproduct of the enzymatic production of nitric oxide from the amino acid arginine, catalyzed by nitric oxide synthase. This is an essential reaction in the body because nitric oxide is an important vasodilator required for regulating blood pressure.

Biosynthesis

Citrulline is made from ornithine and carbamoyl phosphate in one of the central reactions in the urea cycle. It is also produced from arginine as a by-product of the reaction catalyzed by NOS family (NOS; EC 1.14.13.39). It is made from arginine by the enzyme trichohyalin at the inner root sheath and medulla of hair follicles. Arginine is first oxidized into N-hydroxyl-arginine, which is then further oxidized to citrulline concomitant with release of nitric oxide.

Function

Several proteins contain citrulline as a result of a posttranslational modification. These citrulline residues are generated by a family of enzymes called peptidylarginine deiminases (PADs), which convert arginine into citrulline in a process called citrullination or deimination. Proteins that normally contain citrulline residues include myelin basic protein (MBP), filaggrin, and several histone proteins, whereas other proteins, such as fibrin and vimentin are susceptible to citrullination during cell death and tissue inflammation.

Patients with rheumatoid arthritis often have detectable antibodies against proteins containing citrulline. Although the origin of this immune response is not known, detection of antibodies reactive with citrulline (anti-citrullinated protein antibodies) containing proteins or peptides is now becoming an important help in the diagnosis of rheumatoid arthritis.

Circulating citrulline concentration is, in humans, a biomarker of intestinal functionality.

Sources

Citrulline in the form of citrulline malate is sold as a performance-enhancing athletic dietary supplement, which was shown to reduce muscle fatigue in a preliminary human study.

The rind of watermelon (*Citrullus lanatus*) is a natural source of citrulline, discussed in one report as a precursor to producing nitric oxide which is a physiological factor in relaxing vascular smooth muscle and erectile organs.

Domoic Acid

Domoic acid (DA) is a kainic acid analog neurotoxin that causes amnesic shellfish poisoning (ASP). It is produced by algae and accumulates in shellfish, sardines, and anchovies. When sea lions, otters, cetaceans, humans etc., then eat contaminated animals poisoning may result. Exposure to the biotoxin affects the brain, causing seizures, and possibly death.

History

There has been little use of domoic acid throughout history except for in Japan, where it has been used as an anthelmintic for centuries. Domoic acid was first isolated in 1959 from a species of red algae, *Chondria armata*, in Japan; commonly referred to as "doumoi" (in Tokunoshima's dialect word) or "hanayanagi". Poisonings in history have been rare, or undocumented; however, it is thought that the increase in human activities is resulting in an increasing frequency of toxic algal blooms along coastlines in recent years. Consequently, poisonings have been affecting sea animals, birds, and humans.

In 1961, seabirds attacked the Santa Cruz and Capitola areas in California, and though it was never confirmed, they were thought to be under the influence of domoic acid.

In 1987, on Prince Edward Island, Canada, there was a shellfish poisoning resulting in 3 deaths. Blue mussels (*Mytulis edulis*) contaminated with domoic acid were blamed.

An incident in which domoic acid may have been involved took place on June 22, 2006, when a California brown pelican flew through the windshield of a car on the Pacific Coast Highway.

Chemistry

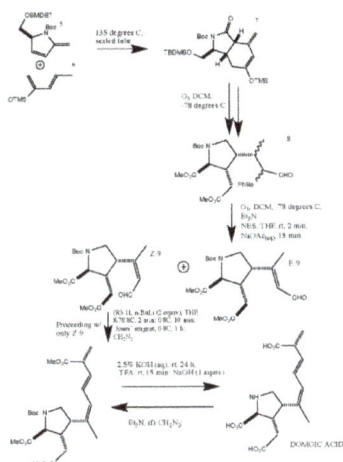

Synthesis of Domoic Acid as described by Jonathan Clayden, Benjamin Read and Katherine R. Hebditch, in their article Chemistry of domoic acid, isodomoic acids, and their analogues

General

Domoic acid is a structural analog of kainic acid and proline. Ohfune and Tomita, who wanted to investigate its absolute stereochemistry, were the first and only to synthesize domoic acid in 1982.

Biosynthesis

In 1999, using [13]C- and ^{14}C-labelled precursors, the biosynthesis of domoic acid in the diatom genus *Pseudo-nitzschia* was examined. After addition of [1,2-13C2]-acetate, NMR spectroscopy showed enrichment of every carbon in domoic acid, indicating incorporation of the carbon isotopes. This enrichment was consistent with two biosynthetic pathways. The labeling pattern determined that domoic acid can be biosynthesized by an isoprenoid intermediate in combination with a tricarboxylic acid (TCA) cycle intermediate.

Synthesis

Using intermediates 5 and 6, a Diels-Alder reaction produced a bicyclic compound (7). 7 then underwent ozonolysis to open the six-membered ring leading to selenide (8). 8 was then deselenated to form 9 (E-9 and Z-9), lastly leading to the formation of (-) domoic acid.

Mechanism of Action

The effects of domoic acid have been attributed to several mechanisms, but the one of concern is through glutamate receptors. Domoic acid is an excitatory amino acid analogue of glutamate; a neurotransmitter in the brain that activates glutamate receptors. Domoic acid has a very strong affinity for these receptors, which results in excitotoxicity initiated by an integrative action on ionotropic glutamate receptors at both sides of the synapse blocking the channel from rapid desensitization. In addition there is a synergistic effect with endogenous glutamate and N-Methyl-D-aspartate receptor agonists that contribute to the excitotoxicity.

In the brain, domoic acid especially damages the hippocampus and amygdaloid nucleus. It damages the neurons by activating AMPA and kainate receptors, causing an influx of calcium. Although calcium flowing into cells is normal, the uncontrolled increase of calcium causes the cells to degenerate. Because the hippocampus may be severely damaged, short-term memory loss occurs. It may also cause kidney damage – even at levels considered safe for human consumption, a new study in mice has revealed. The kidney is affected at a hundred times lower than the concentration allowed under FDA regulations.

Symptoms

Humans	Other Animals
vomiting	head weaving
nausea	seizures
diarrhea and abdominal cramps within 24 hours of ingestion	bulging eyes
headache	mucus from the mouth
dizziness	disorientation
confusion, disorientation	death
loss of short-term memory	
motor weakness	
seizures	
profuse respiratory secretions	
cardiac arrhythmias	
coma and possible death	

Toxicology

Domoic acid producing algal blooms are associated with the phenomenon of amnesic shellfish poisoning (ASP). Domoic acid can bioaccumulate in marine organisms such as shellfish, anchovies, and sardines that feed on the phytoplankton known to produce this toxin. It can accumulate in high concentrations in the tissues of these plankton feeders when the toxic phytoplankton are high in concentration in the surrounding waters. Domoic acid is a neurotoxin that inhibits neurochemical processes, causing short-term memory loss, brain damage, and, in severe cases, death in humans. In marine mammals, domoic acid typically causes seizures and tremors.

Studies have shown that there are no symptomatic effects in humans at levels of 0.5 mg/kg of body weight. In the 1987 domoic acid poisoning on Prince Edward Island concentrations ranging from 0.31–1.28 mg/kg of muscle tissue were noted in people that became ill (three of whom died). Dangerous levels of domoic acid have been calculated based on cases such as the one on Prince Edward island. The exact LD_{50} for humans is unknown; for mice the LD_{50} is 3.6 mg/kg.

New research has found that domoic acid is a heat resistant and very stable toxin, which can damage kidneys at concentrations that are 100 times lower than what causes neurological effects.

Diagnosis and Prevention

In order to be diagnosed and treated if poisoned, domoic acid must first be detected. Methods such as ELISA (enzyme-linked immunosorbent assay), or probe development with PCR (polymerase chain reaction) may be used to detect the toxin or the organism producing this toxin.

There is no known antidote available for domoic acid. Therefore if poisoning occurs, it is advised to go quickly to a hospital. It is important to note that cooking or freezing affected fish or shellfish tissue that are contaminated with domoic acid does not lessen the toxicity.

As a public health concern, the concentration of domoic acid in shellfish and shellfish parts at point of sale should not exceed the current permissible limit of 20 mg/kg tissue. In addition during processing shellfish, it is important to pay attention to environmental condition factors.

Pop Culture

On August 18, 1961, in Capitola and Santa Cruz, California there was an invasion of what people described as chaotic seabirds. These birds were believed to be under the influence of domoic acid, and it inspired a scene in Alfred Hitchcock's feature film *The Birds* (1963).

In addition domoic acid was used to poison a witness in the TV series *Elementary*, episode *The Red Team*.

References

- Belitz, H.-D; Grosch, Werner; Schieberle, Peter (2009-02-27). "Food Chemistry". ISBN 9783540699330.

- Nelson, David L.; Cox, Michael M. (2000). Lehninger Principles of Biochemistry (3rd ed.). Worth Publishers. ISBN 1-57259-153-6.

- Meierhenrich, Uwe J. (2008). Amino acids and the asymmetry of life (1st ed.). Springer. ISBN 978-3-540-76885-2.

- Koyack, M. J.; Cheng, R. P. (2006). "Design and Synthesis of β-Peptides With Biological Activity". Protein Design. 340. pp. 95–109. doi:10.1385/1-59745-116-9:95. ISBN 1-59745-116-9. PMID 16957334.

- The Merck Index: An Encyclopedia of Chemicals, Drugs, and Biologicals (11th ed.), Merck, 1989, ISBN 091191028X , 196.

- Weast, Robert C., ed. (1981). CRC Handbook of Chemistry and Physics (62nd ed.). Boca Raton, FL: CRC Press. p. C-83. ISBN 0-8493-0462-8. .

- Foye, William O.; Lemke, Thomas L. (2007). Foye's Principles of Medicinal Chemistry. David A. Williams. Lippincott Williams & Wilkins. p. 446. ISBN 978-0-7817-6879-5.

- D'haenen, Hugo; den Boer, Johan A. (2002). Biological Psychiatry (digitised online by Google books). Paul Willner. John Wiley and Sons. p. 415. ISBN 978-0-471-49198-9. Retrieved 2008-12-26.

- Purves D, Fitzpatrick D, Hall WC, Augustine GJ, Lamantia AS, eds. (2007). Neuroscience (4th ed.). Sunderland, Mass: Sinauer. pp. 135, box 6D. ISBN 0-87893-697-1.

- Sapse AM (2000). Molecular Orbital Calculations for Amino Acids and Peptides. Birkhäuser. ISBN 978-0-8176-3893-1.

- Roth RJ, Cooper JR, Bloom FE (2003). The Biochemical Basis of Neuropharmacology. Oxford [Oxfordshire]: Oxford University Press. p. 106. ISBN 0-19-514008-7.

- Cox M, Lehninger AL, Nelson DR (2000). Lehninger principles of biochemistry (3rd ed.). New York: Worth Publishers. ISBN 1-57259-153-6.

Chemical Synthesis of Amino Acids

Amino acids can be synthesized by a number of ways. Amino acid synthesis is a set of processes that is produced by other compounds. Peptide synthesis, Arndt-Eistert reaction, Corey-Link reaction, Miller-Urey experiment and Strecker amino acid synthesis are some of the themes elucidated in the following section. The topics discussed in the chapter are of great importance to broaden the existing knowledge on amino acids.

Amino Acid Synthesis

Amino acid synthesis is the set of biochemical processes (metabolic pathways) by which the various amino acids are produced from other compounds. The substrates for these processes are various compounds in the organism's diet or growth media. Not all organisms are able to synthesise all amino acids. Humans are excellent example of this, since humans can only synthesise 11 of the 20 standard amino acids (aka non-essential amino acid), and in time of accelerated growth, histidine, can be considered an essential amino acid.

A fundamental problem for biological systems is to obtain nitrogen in an easily usable form. This problem is solved by certain microorganisms capable of reducing the inert $N \equiv N$ molecule (nitrogen gas) to two molecules of ammonia in one of the most remarkable reactions in biochemistry. Ammonia is the source of nitrogen for all the amino acids. The carbon backbones come from the glycolytic pathway, the pentose phosphate pathway, or the citric acid cycle.

In amino acid production, one encounters an important problem in biosynthesis, namely stereochemical control. Because all amino acids except glycine are chiral, biosynthetic pathways must generate the correct isomer with high fidelity. In each of the 19 pathways for the generation of chiral amino acids, the stereochemistry at the α-carbon atom is established by a transamination reaction that involves pyridoxal phosphate. Almost all the transaminases that catalyze these reactions descend from a common ancestor, illustrating once again that effective solutions to biochemical problems are retained throughout evolution.

Biosynthetic pathways are often highly regulated such that building-blocks are synthesized only when supplies are low. Very often, a high concentration of the final product of a pathway inhibits the activity of enzymes that function early in the pathway. Often present are allosteric enzymes capable of sensing and responding to concentrations of regulatory species. These enzymes are similar in functional properties to aspartate transcarbamoylase and its regulators. Feedback and allosteric mechanisms ensure that all twenty amino acids are maintained in sufficient amounts for protein synthesis and other processes.

Nitrogen Fixation

Microorganisms use ATP and reduced ferredoxin, a powerful reductant, to reduce atmospheric nitrogen (N_2) to ammonia (NH_3). An iron-molybdenum cluster in nitrogenase deftly catalyzes the fixation of N_2, a very inert molecule. Higher organisms consume the fixed nitrogen to synthesize amino acids, nucleotides, and other nitrogen-containing biomolecules. The major points of entry of ammonia into metabolism are glutamine or glutamate.

Transamination

Most amino acids are synthesized from α-ketoacids, and later transaminated from another amino acid, usually glutamate. The enzyme involved in this reaction is an aminotransferase.

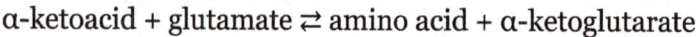

α-ketoacid + glutamate ⇄ amino acid + α-ketoglutarate

Glutamate itself is formed by amination of α-ketoglutarate:

α-ketoglutarate + NH+4 ⇄ glutamate

From Intermediates of the Citric Acid Cycle and other Pathways

Of the basic set of twenty amino acids (not counting selenocysteine), there are eight that human beings cannot synthesize. In addition, the amino acids arginine, cysteine, glycine, glutamine, histidine, proline, serine, and tyrosine are considered conditionally essential, meaning they are not normally required in the diet, but must be supplied exogenously to specific populations that do not synthesize it in adequate amounts. For example, enough arginine is synthesized by the urea cycle to meet the needs of an adult but perhaps not those of a growing child. Amino acids that must be obtained from the diet are called essential amino acids. Nonessential amino acids are produced in the body. The pathways for the synthesis of nonessential amino acids are quite simple. Glutamate dehydrogenase catalyzes the reductive amination of α-ketoglutarate to glutamate. A transamination reaction takes place in the synthesis of most amino acids. At this step, the chirality of the amino acid is established. Alanine and aspartate are synthesized by the transamination of pyruvate and oxaloacetate, respectively. Glutamine is synthesized from NH4+ and glutamate, and asparagine is synthesized similarly. Proline and arginine are derived from glutamate. Serine, formed from 3-phosphoglycerate, is the precursor of glycine and cysteine. Tyrosine is synthesized by the hydroxylation of phenylalanine, an essential amino acid. The pathways for the biosynthesis of essential amino acids are much more complex than those for the nonessential ones. Activated Tetrahydrofolate, a carrier of one-carbon units, plays an important role in the metabolism of amino acids and nucleotides. This coenzyme carries one-carbon units at three oxidation states, which are interconvertible: most reduced—methyl; intermediate—methylene; and most oxidized—formyl, formimino, and methenyl. The major donor of activated methyl groups is S-adenosylmethionine, which is synthesized by the transfer of an adenosyl group from ATP to the sulfur atom of methionine. S-Adenosylhomocysteine is formed when the activated methyl group is transferred to an acceptor. It is hydrolyzed to adenosine and homocysteine, the latter of which is then methylated to methionine to complete the activated methyl cycle.

Cortisol inhibits protein synthesis.

Regulation by Feedback Inhibition

Most of the pathways of amino acid biosynthesis are regulated by feedback inhibition, in which the committed step is allosterically inhibited by the final product. Branched pathways require extensive interaction among the branches that includes both negative and positive regulation. The regulation of glutamine synthetase from *E. coli* is a striking demonstration of cumulative feedback inhibition and of control by a cascade of reversible covalent modifications.

A-Ketoglutarates

The α-ketoglutarate family of amino acid synthesis (synthesis of glutamate, glutamine, proline and arginine) begins with α-ketoglutarate, an intermediate in the Citric Acid Cycle. The concentration of α-ketoglutarate is dependent on the activity and metabolism within the cell along with the regulation of enzymatic activity. In *E. coli* citrate synthase, the enzyme involved in the condensation reaction initiating the Citric Acid Cycle is strongly inhibited by α-ketoglutarate feedback inhibition and can be inhibited by DPNH as well high concentrations of ATP. This is one of the initial regulations of the α-ketoglutarate family of amino acid synthesis.

The regulation of the synthesis of glutamate from α-ketoglutarate is subject to regulatory control of the Citric Acid Cycle as well as mass action dependent on the concentrations of reactants involved due to the reversible nature of the transamination and glutamate dehydrogenase reactions.

The conversion of glutamate to glutamine is regulated by glutamine synthetase (GS) and is a highly significant step in nitrogen metabolism. This enzyme is regulated by at least four different mechanisms: 1. Repression and depression due to nitrogen levels; 2. Activation and inactivation due to enzymatic forms (taut and relaxed); 3. Cumulative feedback inhibition through end product metabolites; and 4. Alterations of the enzyme due to adenylation and deadenylation. In rich nitrogenous media or growth conditions containing high quantities of ammonia there is a low level of GS, whereas in limiting quantities of ammonia the specific activity of the enzyme is 20-fold higher. The confirmation of the enzyme plays a role in regulation depending on if GS is in the taut or relaxed form. The taut form of GS is fully active but, the removal of manganese converts the enzyme to the relaxed state. The specific conformational state occurs based on the binding of specific divalent cations and is also related to adenylation. The feedback inhibition of GS is due to a cumulative feedback due to several metabolites including L-tryptophan, L-histidine, AMP, CTP, glucosamine-6-phosphate and carbamyl phosphate, alanine, and glycine. An excess of any one product does not individually inhibit the enzyme but a combination or accumulation of all the end products have a strong inhibitory effect on the synthesis of glutamine. Glutamine synthase activity is also inhibited via adenylation. The adenylation activity is catalyzed by the bifunctional adenylyltransferase/adenylyl removal (AT/AR) enzyme. Glutamine and a regulatory protein called PII act together to stimulate adenylation.

The regulation of proline biosynthesis can be dependent on the initial controlling step through negative feedback inhibition. In *E. coli*, proline allosterically inhibits Glutamate 5-kinase which catalyzes the reaction from L-glutamate to an unstable intermediate L-γ-Glutamyl phosphate.

Arginine synthesis also utilizes negative feedback as well as repression through a repressor encoded by the gene *argR*. The gene product of *argR*, ArgR an aporepressor, and arginine as a corepressor affect the operon of arginine biosynthesis. The degree of repression is determined by the concentrations of the repressor protein and corepressor level.

Erythrose 4-phosphate and Phosphoenolpyruvate

Phenylalanine, tyrosine, and tryptophan are known as the aromatic amino acids. The synthesis of all three share a common beginning to their pathways; the formation of chorismate from phosphoenolpyruvate (PEP) and erythrose 4- phosphate (E4P). The first step, condensation of 3-deoxy-D-arabino-heptulosonic acid 7-phosphate (DAHP) from PEP/E4P, uses three isoenzymes AroF, AroG, and AroH. Each one of these has its synthesis regulated from tyrosine, phenylalanine, and tryptophan, respectively. These isoenzymes all have the ability to help regulate synthesis of DAHP by the method of feedback inhibition. This acts in the cell by monitoring the concentrations of each of the three aromatic amino acids. When there is too much of any one of them, that one will allosterically control the DAHP synthetase by "turning it off". With the first step of the common pathway shut off, synthesis of the three amino acids can not proceed. The rest of the enzymes in the common pathway (conversion of DAHP to chorismate) appear to be synthesized constitutively, except for shikimate kinase which can be inhibited by shikimate through linear mixed-type inhibition. If too much shikimate has been produced then it can bind to shikimate kinase to stop further production.

Besides the regulations described above, each amino acids terminal pathway can be regulated. These terminal pathways progress from chorismate to the final end product, either tyrosine, phenylalanine, or tryptophan. Each one of these pathways is regulated in a similar fashion to the common pathway; with feedback inhibition on the first committed step of the pathway.

Tyrosine and phenylalanine share the same initial step in their terminal pathways, chorismate converted to prephenate which is converted to an amino acid-specific intermediate. This process is mediated by a phenylalanine (PheA) or tyrosine (TyrA) specific chorismate mutase-prephenate dehydrogenase. The reason for the amino acid-specific enzymes is because PheA uses a simple dehydrogenase to convert prephenate to phenylpyruvate, while TyrA uses a NAD-dependent dehydrogenase to make 4-hydroxylphenylpyruvate. Both PheA and TyrA are feedback inhibited by their respective amino acids. Tyrosine can also be inhibited at the transcriptional level by the TyrR repressor. TyrR binds to the TyrR boxes on the operon near the promoter of the gene that it wants to repress.

In the terminal-tryptophan synthesis pathway, the initial step converts chorismate to anthranilate using anthranilate synthase. This enzyme requires either ammonia or glutamine as the amino group donor. Anthranilate synthase is regulated by the gene products of trpE and trpD. trpE encodes the first subunit, which binds to chorismate and moves the amino group from the donor to chorismate. trpD encodes the second subunit, which is simply used to bind glutamine and use it as the amino group donor so that the amine group can transfer to the chorismate. Anthranilate synthase is also regulated by feedback inhibition. The finished product of tryptophan, once produced in great enough quantities, is able to act as the co-repressor to the TrpR repressor which represses expression of the trp operon.

Oxaloacetate/Aspartate

The oxaloacetate/aspartate family of amino acids is composed of lysine, asparagine, methionine, threonine, and isoleucine. Aspartate can be converted into lysine, asparagine, methionine and threonine. Threonine also gives rise to isoleucine. All of these amino acids contain different mechanisms for their regulation, some being more complex than others. All the enzymes in this biosynthetic pathway are subject to regulation via feedback inhibition and/or repression at the genetic level. As is typical in highly branched metabolic pathways, there is additional regulation at each branch point of the pathway. This type of regulatory scheme allows control over the total flux of the aspartate pathway in addition to the total flux of individual amino acids. The aspartate pathway uses L-aspartic acid as the precursor for the biosynthesis of one fourth of the building block amino acids. Without this pathway, protein synthesis would not be possible.

Aspartate

The enzyme aspartokinase, which catalyzes the phosphorylation of aspartate and initiates its conversion into other amino acids, can be broken up into 3 isozymes, AK-I, II and III. AK-I is feed-back inhibited by threonine, while AK-II and III are inhibited by lysine. As a sidenote, AK-III catalyzes the phosphorylation of aspartic acid that is the commitment step in this biosynthetic pathway. The higher the concentration of threonine or lysine, the more aspartate kinase becomes downregulated.

Lysine

Lysine is synthesized from aspartate via the diaminopimelate (DAP) pathway. The initial two stages of the DAP pathway are catalyzed by aspartokinase and aspartate semi-aldehyde dehydrogenase and play a key role in the biosynthesis of lysine, threonine and methionine. There are two bifunctional aspartokinase/homoserine dehydrogenases, ThrA and MetL, in addition to a monofunctional aspartokinase, LysC. Transcription of aspartokinase genes is regulated by concentrations of the subsequently produced amino acids, lysine, threonine and methionine. The higher these amino acids concentrations, the less the gene is transcribed. ThrA and LysC are also feed-back inhibited by threonine and lysine. Finally, DAP decarboxylase LysA mediates the last step of the lysine synthesis and is common for all studied bacterial species. The formation of aspartate kinase (AK), which catalyzes the phosphorylation of aspartate and initiates its conversion into other amino acids, is also inhibited by both lysine and threonine, which prevents the formation of the amino acids derived from aspartate. Additionally, high lysine concentrations inhibit the activity of dihydrodipicolinate synthase (DHPS). So, in addition to inhibiting the first enzyme of the aspartate families biosynthetic pathway, lysine also inhibits the activity of the first enzyme after the branch point, i.e. the enzyme that is specific for lysine's own synthesis.

Asparagine

There are two different asparagine synthetases found in bacterial species. These two synthetases, which are both referred to as the AsnC protein, are coded for by two genes: AsnA and AsnB. AsnC is autogenously regulated, which is where the product of a structural gene regulates the expression of the operon in which the genes reside. The stimulating effect of AsnC on AsnA transcription is downregulated by asparagine. However, the autoregulation of AsnC is not affected by asparagine.

Methionine

Methionine synthesis is under tight regulation. The repressor protein MetJ, in cooperation with the corepressor protein S-adenosyl-methionine, mediates the repression of methionine's biosynthetic pathway. Recently, a new regulator focus, MetR has been identified. The MetR protein is required for MetE and MetH gene expression and functions as a transactivator of transcription for these genes. MetR transcriptional activity is regulated by homocystein, which is the metabolic precursor of methionine. It is also known that vitamin B12 can repress MetE gene expression, which is mediated by the MetH holoenzyme.

Threonine

The biosynthesis of threonine is regulated via allosteric regulation of its precursor, homoserine, by structurally altering the enzyme homoserine dehydrogenase. This reaction occurs at a key branch point in the pathway, with the substrate homoserine serving as the precursor for the biosynthesis of lysine, methionine, threonin and isoleucine. High levels of threonine result in low levels of homoserine synthesis. The synthesis of aspartate kinase (AK), which catalyzes the phosphorylation of aspartate and initiates its conversion into other amino acids, is feed-back inhibited by lysine, isoleucine, and threonine, which prevents the synthesis of the amino acids derived from aspartate. So, in addition to inhibiting the first enzyme of the aspartate families biosynthetic pathway, threonine also inhibits the activity of the first enzyme after the branch point, i.e. the enzyme that is specific for threonine's own synthesis.

Isoleucine

The enzymes threonine deaminase, dihydroxy acid dehydrase and transaminase are controlled by end-product regulation. I.e. the presence of isoleucine will downregulate the formation of all three enzymes, resulting in the downregulation of threonine biosynthesis. High concentrations of isoleucine also result in the downregulation of aspartate's conversion into the aspartyl-phosphate intermediate, hence halting further biosynthesis of lysine, methionine, threonine, and isoleucine.

Ribose 5-Phosphates

The synthesis of histidine in "E. coli" is a complex pathway involving 10 reactions and 10 enzymes. Synthesis begins with 5-phosphoribosyl-pyrophosphate (PRPP) and finishes with histidine and occurs through the reactions of the following enzymes:

HisG-> HisE/HisI-> HisA-> HisH-> HisF-> HisB-> HisC-> HisB-> HisD (HisE/I and HisB are both bifunctional enzymes)

All of the enzymes are coded for on the his operon. This operon has a distinct block of the leader sequence, called block 1:

Met-Thr-Arg-Val-Gln-Phe-Lys-His-His-His-His-His-His-His-Pro-Asp

This leader sequence is very important for the regulation of histidine in "E. coli". The *his* operon operates under a system of coordinated regulation where all the gene products will be repressed or depressed equally. The main factor in the repression or derepression of histidine synthesis is the concentration of histidine charged tRNAs. The regulation of histidine is actually quite simple considering the complexity of its biosynthesis pathway and, it closely resembles regulation of tryptophan. In this system the full leader sequence has 4 blocks of complementary strands that can form hairpin loops structures. Block one, shown above, is the key to regulation. When histidine charged tRNA levels are low in the cell the ribosome will stall at the string of His residues in block 1. This stalling of the ribosome will allow complementary strands 2 and 3 to form a hairpin loop. The loop formed by strands 2 and 3 forms an anti-terminator and translation of the *his* genes will continue and histidine will be produced. However when histidine charged tRNA levels are high the ribosome will not stall at block 1, this will not allow strands 2 and 3 to form a hairpin. Instead strands 3 and 4 will form a hairpin loop further downstream of the ribosome. The hairpin loop formed by strands 3 and 4 is a terminating loop, when the ribosome comes into contact with the loop, it will be "knocked off" the transcript. When the ribosome is removed the *his* genes will not be translated and histidine will not be produced by the cell.

3-Phosphoglycerates

Serine

Serine is the first amino acid in this family to be produced; it is then modified to produce both glycine and cysteine (and many other biologically important molecules). Serine is formed from 3-phosphoglycerate in the following pathway:

3-phosphoglycerate-> phosphohydroxyl-pyruvate-> phosphoserine-> serine

The conversion from 3-phosphoglycerate to phosphohydroxyl-pyruvate is achieved by the enzyme phosphoglycerate dehydrogenase. This enzyme is the key regulatory step in this pathway. Phosphoglycerate dehydrogenase is regulated by the concentration of serine in the cell. At high concentrations this enzyme will be inactive and serine will not be produced. At low concentrations of serine the enzyme will be fully active and serine will be produced by the bacterium. Since serine is the first amino acid produced in this family both glycine and cysteine will be regulated by the available concentration of serine in the cell.

Glycine

Glycine is synthesized from serine using the enzyme serine hydromethyltransferase (SHMT), which is coded by the gene *glyA*. The enzyme effectively removes a hydroxyl group from serine and replaces it with a methyl group to yield glycine. This reaction is the only way *E. coli* can produce glycine. The regulation of *glyA* is very complex and is known to incorporate serine, glycine, methionine, purines, thymine, and folates however, the full mechanism has yet to be elucidated. The methionine gene product MetR and the methionine intermediate homocysteine are known to positively regulate glyA. Homocysteine is a coactivator of *glyA* and must act in concert with MetR. On the other hand, PurR, a protein which plays a role in purine synthesis and S-adeno-sylme-thionine are known to down regulate *glyA*. PurR binds directly to the control region of *glyA* and effectively turns the gene off so that glycine will not be produced by the bacterium.

Cysteine

Cysteine is a very important molecule for a bacterium's survival. This amino acid harbors a sulfur atom and can actively participate in disulfide bond formation. The genes required for the synthesis of cysteine are coded for on the *cys* regulon. The integration of sulfur into the molecule is positively regulated by CysB. CysB is the main focus of cysteine regulation. Effective inducers of this regulon are N-acetyl-serine (NAS) and very small amounts of reduced sulfur. CysB functions by binding to DNA half sites on the *cys* regulon. These half sites differ in quantity and arrangement depending on the promoter of interest. There is however one half site that is conserved. It lies just upstream of the -35 site of the promoter. There are also multiple accessory sites depending on the promoter. In the absence of the inducer, NAS, CysB will bind the DNA and cover many of the accessory half sites. Without the accessory half sites the regulon cannot be transcribed and cysteine will not be produced. It is believed that the presence of NAS causes CysB to undergo a conformational change. This conformational change allows CysB to bind properly to all the half sites and causes the recruitment of the RNA polymerase. The RNA polymerase will then transcribe the *cys* regulon and cysteine will be produced.

Further regulation is required for this pathway, however. CysB can actually down regulate its own transcription by binding to its own DNA sequence and blocking the RNA polymerase. In this case NAS will act to disallow the binding of CysB to its own DNA sequence. OAS is a precursor of NAS, cysteine itself can inhibit CysE which functions to create OAS. Without the necessary OAS, NAS will not be produced and cysteine will not be produced. There are two other negative regulators of cysteine. These are the molecules sulfide and thiosulfate, they act to bind to CysB and they compete with NAS for the binding of CysB.

Pyruvates

Pyruvate is the end result of glycolysis and can feed into both the TCA cycle and fermentation processes. Reactions beginning with either one or two molecules of pyruvate cause the synthesis of alanine, valine, and leucine. Feedback inhibition of final products is the main method of inhibition, and, in *E. coli*, the *ilvEDA* operon also plays a part in this regulation.

Alanine

Alanine is produced by the transamination of one molecule of pyruvate using two alternate steps: 1) conversion of glutamate to α-ketoglutarate using a glutamate-alanine transaminase, and 2) conversion of valine to α-ketoisovalerate via Transaminase C.

Not much is known about the regulation of alanine synthesis. The only definite method is the bacterium's ability to repress Transaminase C activity by either valine or leucine. Other than that, alanine biosynthesis does not seem to be regulated.

Valine

Valine is produced by a four-enzyme pathway. It begins with the reaction of two pyruvate molecules catalyzed by Acetohydroxy acid synthase yielding α-acetolactate. Step two is the NADPH+ + H+ - dependent reduction of α-acetolactate and migration of the methane groups to produce α, β-dihydroxyisovalerate. This is catalyzed by Acetohydroxy isomeroreductase. The third reaction is the dehydration reaction of α, β-dihydroxyisovalerate catalyzed by Dihydroxy acid dehydrase resulting in α-ketoisovalerate. Finally, a transamination catalyzed either by an alanine-valine transaminase or a glutamate-valine transaminase results in valine.

Valine performs feedback inhibition to inhibit the Acetohydroxy acid synthase used to combine the first two pyruvate molecules.

Leucine

The leucine synthesis pathway diverges from the valine pathway beginning with α-ketoisovalerate. α-Isopropylmalate synthase reacts with this substrate and Acetyl CoA to produce α-isopropylmalate. An isomerase then isomerizes α-isopropylmalate to β-isopropylmalate. The third step is the NAD+-dependent oxidation of β-isopropylmalate via the action of a dehydrogenase to yield α-ketoisocaproate. Finally is the transamination via the action of a glutamate-leucine transaminase to result in leucine.

Leucine, like valine, regulates the first step of its pathway by inhibiting the action of the α-Isopropylmalate synthase. Because leucine is synthesized by a diversion from the valine synthetic pathway, the feedback inhibition of valine on its pathway also can inhibit the synthesis of leucine.

ilvEDA *Operon*

The genes that encode both the Dihydroxy acid dehydrase used in the creation of α-ketoisovalerate and Transaminase E, as well as other enzymes are encoded on the ilvEDA operon. This operon is bound and inactivated by valine, leucine, and isoleucine. (Isoleucine is not a direct derivative of pyruvate, but is produced by the use of many of the same enzymes used to produce valine and, indirectly, leucine.) When one of these amino acids is limited, the gene furthest from the amino-acid binding site of this operon can be transcribed. When a second of these amino acids is limited, the next-closest gene to the binding site can be transcribed, and so forth.

Amino Acids as Precursors to other Biomolecules

Amino acids are precursors of a variety of biomolecules. Glutathione (γ-Glu-Cys-Gly) serves as a sulfhydryl buffer and detoxifying agent. Glutathione peroxidase, a selenoenzyme, catalyzes the reduction of hydrogen peroxide and organic peroxides by glutathione. Nitric oxide, a short-lived messenger, is formed from arginine. Porphyrins are synthesized from glycine and succinyl CoA, which condense to give δ-aminolevuli-nate. Two molecules of this intermediate become linked to form porphobilinogen. Four molecules of porphobilinogen combine to form a linear tetrapyrrole, which cyclizes to uroporphyrinogen III. Oxidation and side-chain modifications lead to the synthesis of protoporphyrin IX, which acquires an iron atom to form heme.

Peptide Synthesis

In organic chemistry, peptide synthesis is the production of peptides, which are organic compounds in which multiple amino acids are linked via amide bonds, also known as peptide bonds. The biological process of producing long peptides (proteins) is known as protein biosynthesis.

Chemistry

Peptides are synthesized by coupling the carboxyl group or C-terminus of one amino acid to the amino group or N-terminus of another. Due to the possibility of unintended reactions, protecting groups are usually necessary. Chemical peptide synthesis starts at the C-terminal end of the peptide and ends at the N-terminus. This is the opposite of protein biosynthesis, which starts at the N-terminal end.

Liquid-phase Synthesis

Liquid-phase peptide synthesis is a classical approach to peptide synthesis. It has been replaced in most labs by solid-phase synthesis. However, it retains usefulness in large-scale production of peptides for industrial purposes.

Solid-phase Synthesis

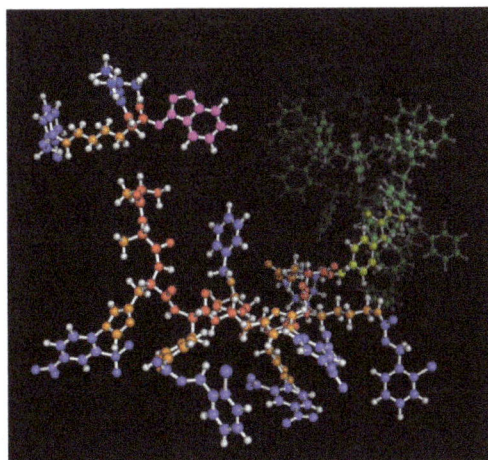

Coupling step in solid-phase peptide synthesis

Solid-phase peptide synthesis (SPPS), pioneered by Robert Bruce Merrifield, caused a paradigm shift within the peptide synthesis community, and it is now the standard method for synthesizing peptides and proteins in the lab. SPPS allows for the synthesis of natural peptides which are difficult to express in bacteria, the incorporation of unnatural amino acids, peptide/protein backbone modification, and the synthesis of D-proteins, which consist of D-amino acids.

Small porous beads are treated with functional units ('linkers') on which peptide chains can be built. The peptide will remain covalently attached to the bead until cleaved from it by a reagent such as anhydrous hydrogen fluoride or trifluoroacetic acid. The peptide is thus 'immobilized' on the solid-phase and can be retained during a filtration process while liquid-phase reagents and by-products of synthesis are flushed away.

The general principle of SPPS is one of repeated cycles of deprotection-wash-coupling-wash. The free N-terminal amine of a solid-phase attached peptide is coupled to a single N-protected amino acid unit. This unit is then deprotected, revealing a new N-terminal amine to which a further amino acid may be attached. The superiority of this technique partially lies in the ability to perform wash cycles after each reaction, removing excess reagent with all of the growing peptide of interest remaining covalently attached to the insoluble resin.

The overwhelmingly important consideration is to generate extremely high yield in each step. For example, if each coupling step were to have 99% yield, a 26-amino acid peptide would be synthesized in 77% final yield (assuming 100% yield in each deprotection); if each step were 95%, it would be synthesized in 25% yield. Thus each amino acid is added in major excess (2~10x) and coupling amino acids together is highly optimized by a series of well-characterized agents.

There are two majorly used forms of SPPS – Fmoc and Boc. Unlike ribosome protein synthesis, solid-phase peptide synthesis proceeds in a C-terminal to N-terminal fashion. The N-termini of amino acid monomers is protected by either of these two groups and added onto a deprotected amino acid chain.

Automated synthesizers are available for both techniques, though many research groups continue to perform SPPS manually.

SPPS is limited by yields, and typically peptides and proteins in the range of 70 amino acids are pushing the limits of synthetic accessibility. Synthetic difficulty also is sequence dependent; typically amyloid peptides and proteins are difficult to make. Longer lengths can be accessed by using native chemical ligation to couple two peptides together with quantitative yields.

Since its introduction over 40 years ago, SPPS has been significantly optimized. First, the resins themselves have been optimized. Furthermore, the 'linkers' between the C-terminal amino acid and polystyrene resin have improved attachment and cleavage to the point of mostly quantitative yields. The evolution of side chain protecting groups has limited the frequency of unwanted side reactions. In addition, the evolution of new activating groups on the carboxyl group of the incoming amino acid have improved coupling and decreased epimerization. Finally, the process itself has been optimized. In Merrifield's initial report, the deprotection of the α-amino group resulted in the formation of a peptide-resin salt, which required neutralization with base prior to coupling. The time between neutralization of the amino group and coupling of the next amino acid allowed for aggregation of peptides, primarily through the formation of secondary structures, and adversely affected coupling. The Kent group showed that concomitant neutralization of the α-amino group and coupling of the next amino acid led to improved coupling. Each of these improvements has helped SPPS become the robust technique that it is today.

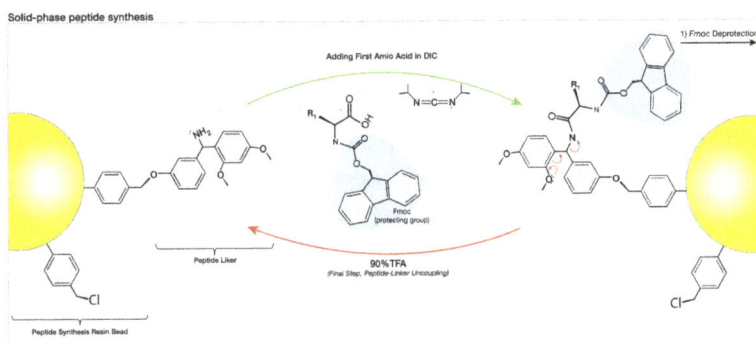

Solid-phase peptide synthesis on a Rink amide resin using Fmoc-α-amine-protected amino acid

BOP SPPS

The use of BOP reagent was first described by Castro et al. in 1975.

Solid Supports

The name solid support implies that reactions are carried out on the surface of the support, but this is not the case. Reactions also occur within these particles, and thus the term "solid support" better describes the insolubility of the polymer. The physical properties of the solid support, and the applications to which it can be utilized, vary with the material from which the support is constructed, the amount of cross-linking, as well as the linker and handle being used. Most scientists in the field believe that supports should have the minimum amount of cross-linking to confer stability. This should result in a well-solvated system where solid-phase peptide synthesis can be carried out. Nonetheless, the characteristics of an efficient solid support include:

- It must be physically stable and permit the rapid filtration of liquids, such as excess reagents

- It must be inert to all reagents and solvents used during SPPS

- It must swell extensively in the solvents used to allow for penetration of the reagents

- It must allow for the attachment of the first amino acid

There are four primary types of solid supports:

1. Gel-type supports: These are highly solvated polymers with an equal distribution of functional groups. This type of support is the most common, and includes:

    - Polystyrene: Styrene cross-linked with 1–2% divinylbenzene

    - Polyacrylamide: A hydrophilic alternative to polystyrene

    - Polyethylene glycol (PEG): PEG-Polystyrene (PEG-PS) is more stable than polystyrene and spaces the site of synthesis from the polymer backbone

    - PEG-based supports: Composed of a PEG-polypropylene glycol network or PEG with polyamide or polystyrene

2. Surface-type supports: Many materials have been developed for surface functionalization, including controlled pore glass, cellulose fibers, and highly cross-linked polystyrene.

3. Composites: Gel-type polymers supported by rigid matrices.

Polystyrene Resin

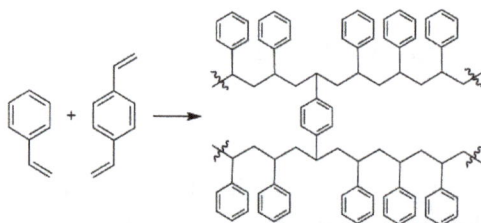

Polystyrene cross-linked with divinylbenzene. This is the most common solid support used in SPPS, and was the support pioneered by R. Bruce Merrifield.

Polystyrene resin is a versatile resin and it is quite useful in multi-well, automated peptide synthesis, due to its minimal swelling in dichloromethane. The initial support used by R. Bruce Merrifield was polysytrene cross-linked with 2% divinylbenzene. This support is sometimes referred to as the 'Merrifield resin.' This resin produces a hydrophobic bead that is solvated by a nonpolar solvent such as dichloromethane or dimethylformamide. Since then, new resins have been developed with the following advantages:

- Enhanced swelling or rigidity (a property of mechanical strength)

- Chemical inertness

Highly cross-linked (50%) polystyrene has been developed that possesses the features of increased mechanical stability, better filtration of reagents and solvents, and rapid reaction kinetics.

Polyamide Resin

Polyamide resin is also a useful and versatile resin. It seems to swell much more than polystyrene, in which case it may not be suitable for some automated synthesizers, if the wells are too small.

PEG Hybrid Polystyrene Resin

An example of this type of resin is the Tentagel resin. The base resin is polystyrene onto which is attached long chains (Mw ca. 3000 Da) of polyethylene glycol (PEG; also known as polyethylene oxide). Synthesis is carried out on the distal end of the PEG spacer making it suited for long and difficult peptides. In addition it is also attractive for the synthesis of combinatorial Peptide libraries and on resin screening experiments. It does not expand much during synthesis making it a preferred resin for robotic peptide synthesis.

PEG-based Resin

ChemMatrix(R) is a new type of resin which is based on PEG that is crosslinked. ChemMatrix(R) has claimed a high chemical and thermal stability (is compatible with Microwave synthesis) and has shown higher degrees of swellings in acetonitrile, dichloromethane, DMF, N-methylpyrrolidone, TFA and water compared to the polystyrene-based resins. ChemMatrix has shown significant improvements to the synthesis of hydrophobic sequences. ChemMatrix may be useful for the synthesis of difficult and long peptides.

Improvements to solid supports used for peptide synthesis enhance their ability to withstand the repeated use of TFA during the deprotection step of SPPS. Furthermore, different resins allow for different functional groups at the C-terminus. The oxymethylphenylacetamidomethyl (PAM) resin results in the conventional C-terminal carboxylic acid. On the other hand, the paramethylbenzhydrylamine (pMBHA) resin yields a C-terminal amide, which is useful in mimicking the interior of a protein.

Along with the development of Fmoc SPPS, different resins have also been created to be removed by TFA. Similar to the Boc strategy, two primary resins are used, based on whether a C-terminal carboxylic acid or amide is desired. The Wang resin is the most commonly used resin for peptides with C-terminal carboxylic acids. If a C-terminal amide is desired, the Rink amide resin is used.

Protecting Groups

Amino acids have reactive moieties at the N- and C-termini, which facilitates amino acid coupling during synthesis. Many amino acids also have reactive side chain functional groups, which can interact with free termini or other side chain groups during synthesis and peptide elongation and negatively influence yield and purity. To facilitate proper amino acid synthesis with minimal side chain reactivity, chemical groups have been developed to bind to specific amino acid functional groups and block, or protect, the functional group from nonspecific reactions. These protecting groups, while vast in nature, can be separated into three groups, as follows:

- N-terminal protecting groups

- C-terminal protecting groups (mostly used in liquid-phase synthesis)

- Side chain protecting groups

Purified, individual amino acids are reacted with these protecting groups prior to synthesis and then selectively removed during specific steps of peptide synthesis.

N-terminal Protecting Groups

Amino acids are added in excess to ensure complete coupling during each synthesis step, and without N-terminal protection, polymerization of unprotected amino acids could occur, resulting in low peptide yield or synthesis failure. N-terminal protection requires an additional step of removing the protecting group, termed deprotection, prior to the coupling step, creating a repeating design flow as follows:

- Protecting group is removed from the trailing amino acids in a deprotection reaction

- Deprotection reagents are washed away to provide a clean coupling environment

- Protected amino acids dissolved in a solvent such as dimethylformamide (DMF) combined with coupling reagents are pumped through the synthesis column

- Coupling reagents are washed away to provide clean deprotection environment

Currently, two protecting groups (t-Boc, Fmoc) are commonly used in solid-phase peptide synthesis. Their lability is caused by the carbamate group which readily releases CO_2 for an irreversible decoupling step.

t-Boc Protecting Group

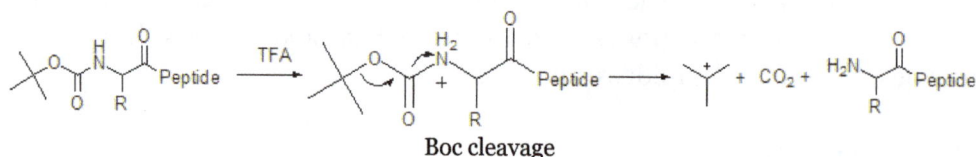

Boc cleavage

The original method for the synthesis of proteins relied on *tert*-butyloxycarbonyl (or more simply "Boc") to temporarily protect the α-amino group. In this method, the Boc group is covalently bound to the amino group to suppress its nucleophilicity. The C-terminal amino acid is covalently linked to the resin through a linker. Next, the Boc group is removed with acid, such as trifluoroacetic acid (TFA). This forms a positively charged amino group (in the presence of excess TFA; note image on the right illustrates neutral amino group), which is neutralized (via in-situ or non-in-situ methods) and coupled to the incoming activated amino acid. Reactions are driven to completion by the use of excess (two- to four-fold) activated amino acid. After each deprotection and coupling step, a wash with dimethylformamide (DMF) is performed to remove excess reagents, allowing for high yields (~99%) during each cycle.

t-Boc protecting strategies retain usefulness in reducing peptide aggregation during synthesis. t-Boc groups can be added to amino acids with t-Boc anhydride and a suitable base. Some researchers prefer Boc SPPS for complex syntheses . In addition, when synthesizing nonnatural peptide analogs, which are base-sensitive (such as depsipeptides), the t-Boc protecting group is necessary, because Fmoc SPPS uses a base to deprotect the α-amino group.

Permanent side-chain protecting groups are typically benzyl or benzyl-based groups. Final removal of the peptide from the linkage occurs simultaneously with side-chain deprotection with anhydrous hydrogen fluoride via hydrolytic cleavage. The final product is a fluoride salt which is relatively easy to solubilize. Importantly, scavengers such as cresol are added to the HF in order to prevent reactive t-butyl cations from generating undesired products. In fact, the use of harsh hydrogen fluoride may degrade some peptides, which was the premise for the development of a milder, base-labile method of SPPS—namely, the Fmoc method.

Fmoc Protecting Group

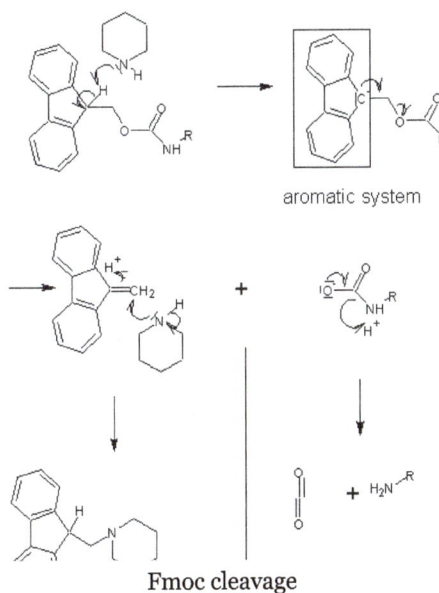

aromatic system

Fmoc cleavage

The capacity for anhydrous hydrogen fluoride to degrade proteins during the final cleavage conditions led to a new α-amino protecting group based on 9-fluorenylmethyloxycarbonyl (Fmoc). The Fmoc method allows for a milder deprotection scheme. This method utilizes a base, usually piperidine (20–50%) in DMF in order to remove the Fmoc group to expose the α-amino group for reaction with an incoming activated amino acid. Unlike the acid used to deprotect the α-amino group in Boc methods, Fmoc SPPS uses a base, and thus the exposed amine is neutral. Therefore, no neutralization of the peptide-resin is required, but the lack of electrostatic repulsions between the peptides can lead to increased aggregation. Because the liberated fluorenyl group is a chromophore, deprotection by Fmoc can be monitored by UV absorbance of the runoff, a strategy which is employed in automated synthesizers.

The advantage of Fmoc is that it is cleaved under very mild basic conditions (e.g. piperidine), but stable under acidic conditions, although this has not always held true in certain synthetic sequences. This allows mild acid-labile protecting groups that are stable under basic conditions, such as Boc and benzyl groups, to be used on the side-chains of amino acid residues of the target peptide. This orthogonal protecting group strategy is common in organic synthesis. Fmoc is preferred over BOC due to ease of cleavage; however it is less atom-economical, as the fluorenyl group is much larger than the tert-butyl group. Accordingly, prices for Fmoc amino acids were high until the large-scale piloting of one of the first synthesized peptide drugs, enfuvirtide, began in the 1990s, when market demand adjusted the relative prices of the two sets of amino acids.

Semipermanent side chain protecting groups are t-butyl-based, and final cleavage of the protein from the resin and removal of permanent protecting groups is performed with TFA in the presence of scavengers. Water and triisopropylsilane (TIPS) present in a 1:1 ratio are often used as scavengers. Thus, the Fmoc method is orthogonal in two directions: deprotection of any α-amino group, deprotection of side groups and final cleavage from the resin occur by independent mechanisms. The resulting final product is a TFA salt, which is more difficult to solubilize than the fluoride salts generated in Boc SPPS. This method is thus milder than the Boc method because the deprotection/cleavage-from-resin steps occur with different conditions rather than with different reaction rates.

Comparison of Boc and Fmoc Solid-phase Peptide Synthesis

Both the Fmoc and Boc methods offer advantages and disadvantages. The selection of one technique over another is thus made on a case-by-case basis.

	Boc	Fmoc
Requires special equipment	Yes	No
Cost of reagents	Lower	Higher
Solubility of peptides	Higher	Lower

Purity of hydrophobic peptides	High	May be lower
Problems with aggregation	Less frequently	More frequently
Synthesis time	~20 min/amino acid	~20–60 min/amino acid
Cleavage from resin	HF	TFA
Safety	Potentially dangerous	Relatively safe
Orthogonal	No	Yes

Boc SPPS uses special equipment to handle the final cleavage and deprotection step, which requires anhydrous hydrogen fluoride. Because the final cleavage of the peptide with Fmoc SPPS uses TFA, this special equipment is not necessary. The solubility of peptides generated by Boc SPPS is generally higher than those generated with the Fmoc method, because fluoride salts are higher in solubility than TFA salts. Next, problems with aggregation are generally more of an issue with Fmoc SPPS. This is primarily because the removal of a Boc group with TFA yields a positively charged α-amino group, whereas the removal of an Fmoc group yields a neutral α-amino group. The steric hindrance of the positively charged α-amino group limits the formation of secondary structure on the resin. Finally, the Fmoc method is considered orthogonal, since α-amino group deprotection is with base, while final cleavage from the resin is with acid. The Boc method utilizes acid for both deprotection and cleavage from the resin. Based on this comparison, one sees that both methods possess advantages and disadvantages. Thus, several factors help to decide which method may be preferable.

DMF must be 'peptide grade' i.e. little/no impurities and must also be 'fresh'. This is due to the fact that DMF undergoes photolysis to form carbon monoxide and dimethylamine. Dimethylamine may remove the Fmoc group and, therefore, lead to impurities.

Benzyloxy-carbonyl (Z) Group

Another carbamate-based group is the benzyloxy-carbonyl (Z) group. It is removed in harsher conditions: HBr/acetic acid or catalytic hydrogenation. The first use of (Z) group as protecting groups was done by Max Bergmann who synthesised oligopeptides. Today it is almost exclusively used for side chain protection.

Alloc Protecting Group

The allyloxycarbonyl (alloc) protecting group is often used to protect a carboxylic acid, hydroxyl, or amino group when an orthogonal deprotection scheme is required. It is sometimes used when conducting on-resin cyclic peptide formation, where the peptide is linked to the resin by a side-chain functional group. The alloc group can be removed using tetrakis(triphenylphosphine)palladium(0) along with a 37:2:1 mixture of methylene chloride, acetic acid, and N-Methylmorpholine (NMM) for 2 hours. The resin must then be carefully washed 0.5% DIPEA in DMF, 3 × 10 ml of 0.5% sodium diethylthiocarbamate in DMF, and then 5 × 10 ml of 1:1 DCM:DMF.

Lithographic Protecting Groups

For special applications like protein microarrays lithographic protecting groups are used. Those groups can be removed through exposure to light.

Side Chain Protecting Groups

Amino acid side chains represent a broad range of functional groups and are sites of nonspecific reactivity during peptide synthesis. Because of this, many different protecting groups are required that are usually based on the benzyl (Bzl) or tert-butyl (tBu) group. The specific protecting groups used during the synthesis of a given peptide vary depending on the peptide sequence and the type of N-terminal protection used. Side chain protecting groups are known as permanent or semipermanent protecting groups, because they can withstand the multiple cycles of chemical treatment during synthesis and are only removed during treatment with strong acids after peptide synthesis is completed.

Protection Schemes

Because multiple protecting groups are normally used during peptide synthesis, these groups must be compatible to allow deprotection of distinct protecting groups while not affecting other protecting groups. Protecting schemes are therefore established to match protecting groups so that deprotection of one protecting group does not affect the binding of the other groups. Because N-terminal deprotection occurs continuously during peptide synthesis, protecting schemes have been established in which the different types of side chain protecting groups (Bzl or tBu) are matched to either Boc or Fmoc, respectively, for optimized deprotection. These protecting schemes also incorporate each of the steps of synthesis and cleavage, as described in the table and in later sections of this page.

Activating Groups

For coupling the peptides the carboxyl group is usually activated. This is important for speeding up the reaction. There are two main types of activating groups: carbodiimides and triazolols. However the use of pentafluorophenyl esters (FDPP, PFPOH) and BOP-Cl are useful for cyclising peptides.

Carbodiimides

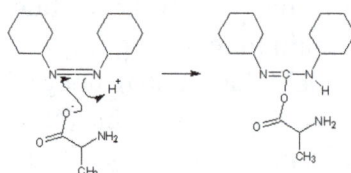

Alanine attaching to DCC

These activating agents were first developed. Most common are dicyclohexylcarbodiimide (DCC) and diisopropylcarbodiimide (DIC). Reaction with a carboxylic acid yields a highly reactive O-acylisourea. During artificial protein synthesis (such as Fmoc solid-state synthesizers), the C-terminus is often used as the attachment site on which the amino acid monomers are added. To enhance the electrophilicity of carboxylate group, the negatively charged oxygen must first be "activated" into a better leaving group. DCC is used for this purpose. The negatively charged oxygen will act as a nucleophile, attacking the central carbon in DCC. DCC is temporarily attached to the former carboxylate group (which is now an ester group), making nucleophilic attack by an amino group (on the attaching amino acid) to the former C-terminus (carbonyl group) more efficient. The problem with carbodiimides is that they are too reactive and that they can therefore cause racemization of the amino acid.

Triazoles

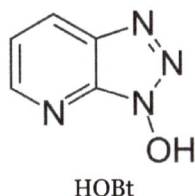

HOBt

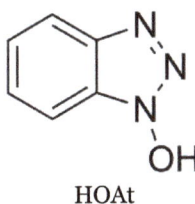

HOAt

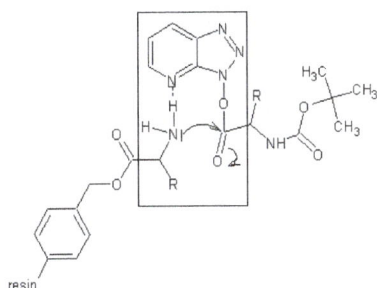

Neighbouring group effect of HOAt

To solve the problem of racemization, triazoles were introduced. The most important ones are 1-hydroxy-benzotriazole (HOBt) and 1-hydroxy-7-aza-benzotriazole (HOAt). Others have been developed. These substances can react with the O-acylurea to form an active ester which is less reactive and less in danger of racemization. HOAt is especially favourable because of a neighbouring group effect. Recently, HOBt has been removed from many chemical vendor catalogues; although almost always found as a hydrate, HOBt may be explosive when allowed to fully dehydrate and shipment by air or sea is heavily restricted. Alternatives to HOBt and HOAt have been introduced. One of the most promising and inexpensive is ethyl 2-cyano-2-(hydroxyimino)acetate (trade name Oxyma Pure), which is not explosive and has a reactivity of that in between HOBt and HOAt.

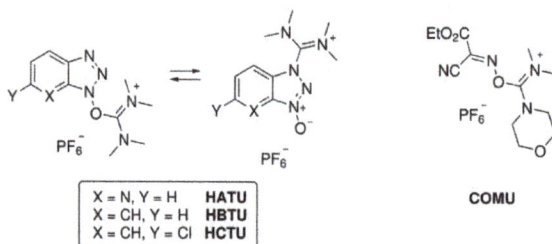

X = N, Y = H	**HATU**
X = CH, Y = H	**HBTU**
X = CH, Y = Cl	**HCTU**

COMU

Uronium-based peptide coupling reagents

Newer developments omit the carbodiimides totally. The active ester is introduced as a uronium or phosphonium salt of a non-nucleophilic anion (tetrafluoroborate or hexafluorophosphate): HBTU, HATU, HCTU, TBTU, PyBOP. Two uronium types of the coupling additive of Oxyma Pure is also available as COMU or TOTU reagent.

Regioselective Disulfide Formation

The formation of multiple native disulfides remains one of the primary challenges of native peptide synthesis by solid-phase methods. Random chain combination typically results in several products with nonnative disulfide bonds. Stepwise formation of disulfide bonds is typically the preferred method, and performed with thiol protecting groups (PGs). Different thiol PGs provide multiple dimensions of orthogonal protection. These orthogonally protected cysteines are incorporated during the solid-phase synthesis of the peptide. Successive removal of these PGs to allow for selective exposure of free thiol groups, leads to disulfide formation in a stepwise manner. The order of removal of these PGs must be considered so that only one group is removed at a time. Using this method, Kiso et al. reported the first total synthesis of insulin by this method in 1993.

The thiol PGs must possess multiple characteristics. First, the PG must be reversible with conditions that do not affect the unprotected side chains. Second, the protecting group must be able to withstand the conditions of solid-phase synthesis. Third, the configuration of the removal of the thiol protecting group must be such that it leaves intact other thiol PGs, if orthogonal protection is desired. That is, the removal of PG A should not affect PG B. Some of the thiol PGs commonly used include the acetamidomethyl (Acm), tert-butyl (But), 3-nitro-2-pyridine sulfenyl (NPYS), 2-pyridine-sulfenyl (Pyr), and triphenylmethyl (Trt) groups. Importantly, the NPYS group can replace the Acm PG to yield an activated thiol.

In the stepwise formation of disulfides to synthesize insulin by Kiso et al., the authors synthesize the A-chain with following protection: CysA6(But); CysA7(Acm);

CysA11(But). Thus, CysA20 is unprotected. Synthesis of the B-chain is performed with the following protection: CysB7(Acm) CysB19(Pyr). The first disulfide bond, CysA20–CysB19, was formed by mixing the two chains in 8 M urea, pH 8 (RT) for 50 min. The second disulfide bond, CysA7–CysB7, was formed by treatment with iodine in aqueous acetic acid to remove the Acm groups. The third disulfide, the intramolecular CysA6–CysA11, was formed by the removal of the But groups by methyltrichlorosilane with diphenyl sulfoxide in TFA. Importantly, formation of the first disulfide in 8 M urea, pH 8 does not affect the other PGs, namely Acm and But groups. Likewise, formation of the second disulfide bond with iodine in aqueous acetic acid does not affect the But groups.

Important to the discussion of disulfide bond formation is the order in which disulfides are formed. From a logical standpoint, the order in which the thiol groups are exposed to form disulfides should be of little consequence, since the other cysteines are protected. Practically, however, the order in which disulfides are formed can have a significant effect on yields. This may be because the formation of the CysA20–CysB19 disulfide may place the thiol group of CysB7 in close proximity with both CysA6 and CysA7, leading to multiple disulfide products. This is one manifestation of the reality that solid-phase peptide synthesis is as much art as it is science.

Synthesizing Long Peptides

Stepwise elongation, in which the amino acids are connected step-by-step in turn, is ideal for small peptides containing between 2 and 100 amino acid residues. Another method is fragment condensation, in which peptide fragments are coupled. Although the former can elongate the peptide chain without racemization, the yield drops if only it is used in the creation of long or highly polar peptides. Fragment condensation is better than stepwise elongation for synthesizing sophisticated long peptides, but its use must be restricted in order to protect against racemization. Fragment condensation is also undesirable since the coupled fragment must be in gross excess, which may be a limitation depending on the length of the fragment.

A new development for producing longer peptide chains is chemical ligation: unprotected peptide chains react chemoselectively in aqueous solution. A first kinetically controlled product rearranges to form the amide bond. The most common form of native chemical ligation uses a peptide thioester that reacts with a terminal cysteine residue. Other methods applicable for covalently linking polypeptides in aqueous solution include the use of split inteins, spontaneous isopeptide bond formation and sortase ligation.

In order to optimize synthesis of long peptides, Zealand Pharma (located in Denmark in Medicon Valley) invented a method for converting a difficult peptide sequence into an easy peptide sequence. The new technology, called SIP-technology, uses "structure-inducing probes" (SIP) to facilitate the synthesis of long peptides. The SIP-technology is a small pre-sequence peptide sequence (e.g. Lysine (Lysn); Glutamic Acid (Glun);

(LysGlu)n) that is incorporated at the C-terminus of subsequent resin bound peptide to induce an alpha-helix-like structure in the peptide. The SIP technology constrains the parent peptide into a more ordered conformation using intramolecular hydrogen bonds. This allows the peptide structure to stabilize, and the utilized hydrogen bonds reduce the likelihood of aggregation and degradation by enzymes. In this way, the SIP technology is designed to optimize peptide synthesis, increase biological half-life, improve peptide stability and inhibit enzymatic degradation without altering pharmacological activity or profile of action.

Coupling Efficiency Vs. Peptide Length

Percentage of peptide product formed after repeated couplings at given efficiency					
Length	Coupling Efficiency	Coupling Efficiency	Coupling Efficiency	Coupling Efficiency	Coupling Efficiency
1	0.995	0.99	0.98	0.97	0.96
5	0.98	0.95	0.92	0.89	0.85
10	0.96	0.91	0.83	0.76	0.69
15	0.93	0.87	0.75	0.65	0.56
20	0.91	0.83	0.68	0.56	0.46
25	0.89	0.79	0.62	0.48	0.38
30	0.86	0.75	0.56	0.41	0.31
35	0.84	0.71	0.50	0.36	0.25
40	0.82	0.67	0.45	0.30	0.20
45	0.80	0.63	0.41	0.26	0.17
50	0.78	0.60	0.37	0.22	0.14
55	0.76	0.58	0.34	0.19	0.11
60	0.74	0.55	0.30	0.17	0.09
65	0.73	0.53	0.27	0.14	0.07
70	0.71	0.50	0.25	0.12	0.06

Microwave Assisted Peptide Synthesis

Although microwave irradiation has been around since the late 1940s, it was not until 1986 that microwave energy was used in organic chemistry. During the end of the 1980s and 1990s, microwave energy was an obvious source for completing chemical reactions in minutes that would otherwise take several hours to days. Through several technical improvements at the end of the 1990s and beginning of the 2000s, micro-wave synthesizers have been designed to provide both low and high energy pockets of microwave energy so that the temperature of the reaction mixture could be controlled. The microwave energy used in peptide synthesis is of a single frequency providing max-imum penetration depth of the sample which is in contrast to conventional kitchen microwaves.

In peptide synthesis, microwave irradiation has been used to complete long peptide sequences with high degrees of yield and low degrees of racemization. Microwave irra-diation during the coupling of amino acids to a growing polypeptide chain is not only catalyzed through the increase in temperature, but also due to the alternating electro-magnetic radiation to which the polar backbone of the polypeptide continuously aligns. Due to this phenomenon, the microwave energy can prevent aggregation and thus in-creases yields of the final peptide product. There is however no clear evidence that mi-crowave is better than simple heating and some peptide laboratories regard microwave just as a convenient method for rapid heating of the peptidyl resin. Heating to above 50–55 degrees Celsius also prevents aggregation and accelerates the coupling.

Despite the main advantages of microwave irradiation of peptide synthesis, the main disadvantage is the racemization which may occur with the coupling of cysteine and histidine. A typical coupling reaction with these amino acids are performed at lower temperatures than the other 18 natural amino acids. A number of peptides do not sur-vive microwave synthesis or heating in general. One of the more serious side effects is dehydration (loss of water) which for certain peptides can be almost quantitative like pancreatic polypeptide (PP). This side effect is also seen by simple heating without the use of microwave.

Cyclic Peptides

On Resin Cyclization

Peptide can be cyclized on solid support. A variety of cylization reagents can be used such as HBTU/HOBt/DIEA, PyBop/DIEA, PyClock/DIEA. Head-to-tail peptides can be made on the solid support. The deprotection of the C-terminus at some suitable point allows on-resin cyclization by amide bond formation with the deprotected N-ter-minus. Once cyclization has taken place, the peptide is cleaved from resin by acidolysis and purified. The strategy for the solid-phase synthesis of cyclic peptides in not limited to attachment through Asp, Glu or Lys side chains. Cysteine has a very reactive sulfhy-dryl group on its side chain. A disulfide bridge is created when a sulfur atom from one Cysteine forms a single covalent bond with another sulfur atom from a second cysteine in a different part of the protein. These bridges help to stabilize proteins, especially those secreted from cells. Some researchers use modified cysteines using S-acetomi-domethyl (Acm) to block the formation of the disulfide bond but preserve the cysteine and the protein's original primary structure.

Arndt–Eistert Reaction

The Arndt–Eistert synthesis is a series of chemical reactions designed to convert a carboxylic acid to a higher carboxylic acid homologue (i.e. contains one additional carbon atom) and is considered a homologation process. Named for the German chemists Fritz Arndt (1885–1969) and Bernd Eistert (1902–1978), Arndt–Eistert synthesis is a popular method of producing β-amino acids from α-amino acids. Acid chlorides react with diazomethane to give diazoketones. In the presence of a nucleophile (water) and a metal catalyst (Ag_2O), diazoketones will form the desired acid homologue.

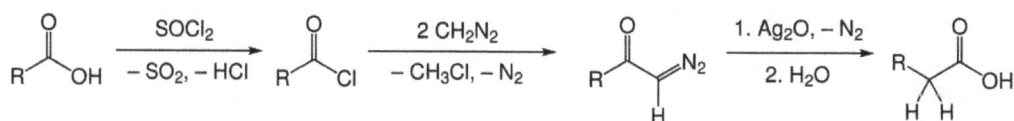

While the classic Arndt–Eistert synthesis uses thionyl chloride to convert the starting acid to an acid chloride, any procedure can be used that will generate an acid chloride.

Diazoketones are typically generated as described here, but other methods such as diazo-group transfer can also apply.

Since diazomethane is toxic and violently explosive, many safer alternatives have been developed, such as the usage of ynolates (Kowalski ester homologation) or diazo(trimethylsilyl)methane.

Reaction Mechanism

The key step in the Arndt–Eistert synthesis is the metal-catalyzed Wolff rearrangement of the diazoketone to form a ketene. In the insertion homologation of *t*-BOC protected (*S*)-phenylalanine (2-amino-3-phenylpropanoic acid), *t*-BOC protected (*S*)-3-amino-4-phenylbutanoic acid is formed.

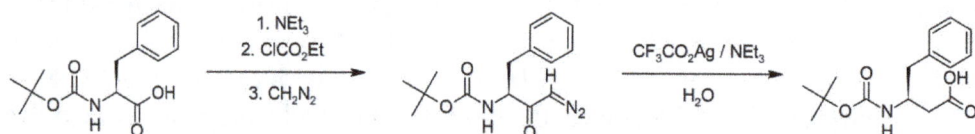

Wolff rearrangement of the α-diazoketone intermediate forms a ketene via a 1,2-rearrangement, which is subsequently hydrolysed to form the carboxylic acid. The consequence of the 1,2-rearrangement is that the methylene group alpha to the carboxyl group in the product is the methylene group from the diazomethane reagant. It has been demonstrated that the rearrangement preserves the stereochemistry of the chiral centre as the product formed from *t*-BOC protected (*S*)-phenylalanine retains the (*S*) stereochemistry with a reported enantiomeric excess of at least 99%.

Heat, light, platinum, silver, and copper salts will also catalyze the Wolff rearrangement to produce the desired acid homologue.

Variations

In the Newman–Beal modification, addition of triethylamine to the diazomethane solution will avoid the formation of α-chloromethylketone side-products.

Corey–Link Reaction

In organic chemistry, the Corey–Link reaction is a name reaction that converts a 1,1,1-tricholoro-2-keto structure into a 2-aminocarboxylic acid (an alpha amino acid) or other acyl functional group with control of the chirality at the alpha position. The reaction is named for E.J. Corey and John Link, who first reported the reaction sequence.

Process

The first stage of the process is the reduction of the carbonyl to give a 1,1,1-tricholoro-2-hydroxy structure. The original protocol used catecholborane with a small amount of one enantiomer of CBS catalyst (a Corey–Itsuno reduction). The choice of chirality of the catalyst thus gives selectivity for one or the other stereochemistry of the alcohol in the product.

This 2-hydroxy structure is then reacted with azide and a nucleophilic base. The multistep reaction mechanism begins with deprotonation of the alcohol, followed by the oxygen-anion attacking the adjacent trichloromethyl position to form an epoxide. The azide then opens this ring by an S_N2 reaction to give a 2-azido structure whose stereochemistry is inverted compared to the original 2-hydroxy. The oxygen, having become attached to the first carbon of the chain during the epoxide formation, simultaneously displaces a second chlorine atom there to form an acyl chloride. An additional nucleophilic reactant, such as hydroxide or an alkoxide, then triggers an acyl substitution there to produce a carboxylic acid or ester. Various other nucleophiles can be used to generate other acyl functional groups. This sequence of steps gives a 2-azido compound, which is then reduced to the 2-amino compound in a separate reaction, typically a Staudinger reaction.

Bargellini Reaction

The Bargellini reaction involves the same type of dichloroepoxy intermediate, formed by a different method, that reacts with a single structure containing two nucleophilic groups. It thus gives products such as morpholinones or piperazinones, alpha-amino esters or amides in which the amine is tethered to the acyl substituent group.

Miller–Urey Experiment

The Miller–Urey experiment (or Miller experiment) was a chemical experiment that simulated the conditions thought at the time to be present on the early Earth, and tested the chemical origin of life under those conditions. The experiment confirmed

Alexander Oparin's and J. B. S. Haldane's hypothesis that putative conditions on the primitive Earth favoured chemical reactions that synthesized more complex organic compounds from simpler inorganic precursors. Considered to be the classic experiment investigating abiogenesis, it was conducted in 1952 by Stanley Miller, with assistance from Harold Urey, at the University of Chicago and later the University of California, San Diego and published the following year.

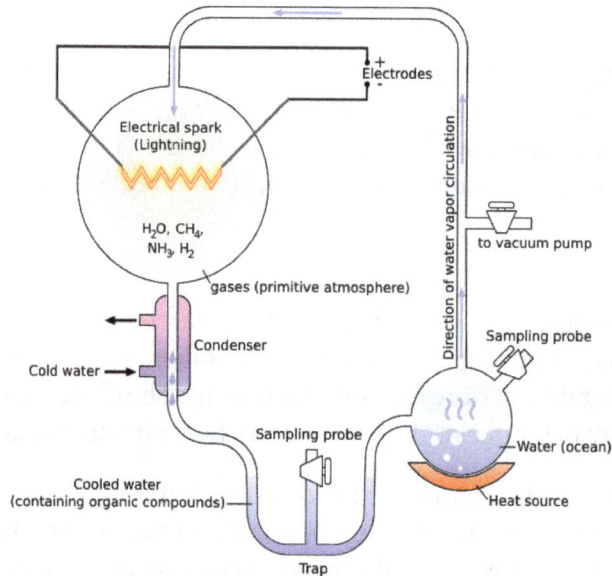

The experiment

After Miller's death in 2007, scientists examining sealed vials preserved from the original experiments were able to show that there were actually well over 20 different amino acids produced in Miller's original experiments. That is considerably more than what Miller originally reported, and more than the 20 that naturally occur in life. More-recent evidence suggests that Earth's original atmosphere might have had a different composition from the gas used in the Miller experiment. But prebiotic experiments continue to produce racemic mixtures of simple to complex compounds under varying conditions.

Experiment

The experiment used water (H_2O), methane (CH_4), ammonia (NH_3), and hydrogen (H_2). The chemicals were all sealed inside a sterile 5-liter glass flask connected to a 500 ml flask half-full of liquid water. The liquid water in the smaller flask was heated to induce evaporation, and the water vapour was allowed to enter the larger flask. Continuous electrical sparks were fired between the electrodes to simulate lightning in the water vapour and gaseous mixture, and then the simulated atmosphere was cooled again so that the water condensed and trickled into a U-shaped trap at the bottom of the apparatus.

Descriptive video of the experiment

After a day, the solution collected at the trap had turned pink in colour. At the end of one week of continuous operation, the boiling flask was removed, and mercuric chloride was added to prevent microbial contamination. The reaction was stopped by adding barium hydroxide and sulfuric acid, and evaporated to remove impurities. Using paper chromatography, Miller identified five amino acids present in the solution: glycine, α-alanine and β-alanine were positively identified, while aspartic acid and α-aminobutyric acid (AABA) were less certain, due to the spots being faint.

In a 1996 interview, Stanley Miller recollected his lifelong experiments following his original work and stated: "Just turning on the spark in a basic pre-biotic experiment will yield 11 out of 20 amino acids."

As observed in all subsequent experiments, both left-handed (L) and right-handed (D) optical isomers were created in a racemic mixture. In biological systems, almost all of the compounds are non-racemic, or homochiral.

The original experiment remains today under the care of Miller and Urey's former student Jeffrey Bada, a professor at UCSD, at the University of California, San Diego, Scripps Institution of Oceanography. The apparatus used to conduct the experiment is on display at the Denver Museum of Nature and Science.

Chemistry of Experiment

One-step reactions among the mixture components can produce hydrogen cyanide (HCN), formaldehyde (CH_2O), and other active intermediate compounds (acetylene, cyanoacetylene, etc.):

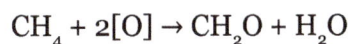

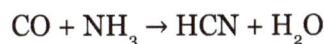

$CO_2 \rightarrow CO + [O]$ (atomic oxygen)

$CH_4 + 2[O] \rightarrow CH_2O + H_2O$

$CO + NH_3 \rightarrow HCN + H_2O$

$CH_4 + NH_3 \rightarrow HCN + 3H_2$ (BMA process)

The formaldehyde, ammonia, and HCN then react by Strecker synthesis to form amino acids and other biomolecules:

$$CH_2O + HCN + NH_3 \rightarrow NH_2\text{-}CH_2\text{-}CN + H_2O$$

$$NH_2\text{-}CH_2\text{-}CN + 2H_2O \rightarrow NH_3 + NH_2\text{-}CH_2\text{-}COOH \text{ (glycine)}$$

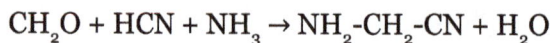

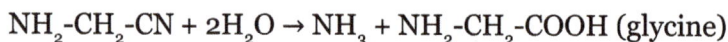

Furthermore, water and formaldehyde can react, via Butlerov's reaction to produce various sugars like ribose.

The experiments showed that simple organic compounds of building blocks of proteins and other macromolecules can be formed from gases with the addition of energy.

Other Experiments

This experiment inspired many others. In 1961, Joan Oró found that the nucleotide base adenine could be made from hydrogen cyanide (HCN) and ammonia in a water solution. His experiment produced a large amount of adenine, the molecules of which were formed from 5 molecules of HCN. Also, many amino acids are formed from HCN and ammonia under these conditions. Experiments conducted later showed that the other RNA and DNA nucleobases could be obtained through simulated prebiotic chemistry with a reducing atmosphere.

There also had been similar electric discharge experiments related to the origin of life contemporaneous with Miller–Urey. An article in *The New York Times* (March 8, 1953:E9), titled "Looking Back Two Billion Years" describes the work of Wollman (William) M. MacNevin at The Ohio State University, before the Miller *Science* paper was published in May 1953. MacNevin was passing 100,000 volt sparks through methane and water vapor and produced "resinous solids" that were "too complex for analysis." The article describes other early earth experiments being done by MacNevin. It is not clear if he ever published any of these results in the primary scientific literature.

K. A. Wilde submitted a paper to *Science* on December 15, 1952, before Miller submitted his paper to the same journal on February 10, 1953. Wilde's paper was published on July 10, 1953. Wilde used voltages up to only 600 V on a binary mixture of carbon dioxide (CO_2) and water in a flow system. He observed only small amounts of carbon dioxide reduction to carbon monoxide, and no other significant reduction products or newly formed carbon compounds. Other researchers were studying UV-photolysis of water vapor with carbon monoxide. They have found that various alcohols, aldehydes and organic acids were synthesized in reaction mixture.

More recent experiments by chemists Jeffrey Bada, one of Miller's graduate students, and Jim Cleaves at Scripps Institution of Oceanography of the University of California, San Diego were similar to those performed by Miller. However, Bada noted that in current models of early Earth conditions, carbon dioxide and nitrogen (N_2) create nitrites, which destroy amino acids as fast as they form. When Bada performed the Miller-type experiment with the addition of iron and carbonate minerals, the products were rich in amino acids. This suggests the origin of significant amounts of amino acids may have occurred on Earth even with an atmosphere containing carbon dioxide and nitrogen.

Earth's Early Atmosphere

Some evidence suggests that Earth's original atmosphere might have contained fewer of the reducing molecules than was thought at the time of the Miller–Urey experiment. There is abundant evidence of major volcanic eruptions 4 billion years ago, which would have released carbon dioxide, nitrogen, hydrogen sulfide (H_2S), and sulfur dioxide (SO_2) into the atmosphere. Experiments using these gases in addition to the ones in the original Miller–Urey experiment have produced more diverse molecules. The experiment created a mixture that was racemic (containing both L and D enantiomers) and experiments since have shown that "in the lab the two versions are equally likely to appear"; however, in nature, L amino acids dominate. Later experiments have confirmed disproportionate amounts of L or D oriented enantiomers are possible.

Originally it was thought that the primitive secondary atmosphere contained mostly ammonia and methane. However, it is likely that most of the atmospheric carbon was CO_2 with perhaps some CO and the nitrogen mostly N_2. In practice gas mixtures containing CO, CO_2, N_2, etc. give much the same products as those containing CH_4 and NH_3 so long as there is no O_2. The hydrogen atoms come mostly from water vapor. In fact, in order to generate aromatic amino acids under primitive earth conditions it is necessary to use less hydrogen-rich gaseous mixtures. Most of the natural amino acids, hydroxyacids, purines, pyrimidines, and sugars have been made in variants of the Miller experiment.

More recent results may question these conclusions. The University of Waterloo and University of Colorado conducted simulations in 2005 that indicated that the early atmosphere of Earth could have contained up to 40 percent hydrogen—implying a much more hospitable environment for the formation of prebiotic organic molecules. The escape of hydrogen from Earth's atmosphere into space may have occurred at only one percent of the rate previously believed based on revised estimates of the upper atmosphere's temperature. One of the authors, Owen Toon notes: "In this new scenario, organics can be produced efficiently in the early atmosphere, leading us back to the organic-rich soup-in-the-ocean concept... I think this study makes the experiments by Miller and others relevant again." Outgassing calculations using a chondritic model for the early earth complement the Waterloo/Colorado results in re-establishing the importance of the Miller–Urey experiment.

In contrast to the general notion of early earth's reducing atmosphere, researchers at the Rensselaer Polytechnic Institute in New York reported the possibility of oxygen available around 4.3 billion years ago. Their study reported in 2011 on the assessment of Hadean zircons from the earth's interior (magma) indicated the presence of oxygen traces similar to modern-day lavas. This study suggests that oxygen could have been released in the earth's atmosphere earlier than generally believed.

Extraterrestrial Sources

Conditions similar to those of the Miller–Urey experiments are present in other regions of the solar system, often substituting ultraviolet light for lightning as the energy source for chemical reactions. The Murchison meteorite that fell near Murchison, Victoria, Australia in 1969 was found to contain over 90 different amino acids, nineteen of which are found in Earth life. Comets and other icy outer-solar-system bodies are thought to contain large amounts of complex carbon compounds (such as tholins) formed by these processes, darkening surfaces of these bodies. The early Earth was bombarded heavily by comets, possibly providing a large supply of complex organic molecules along with the water and other volatiles they contributed. This has been used to infer an origin of life outside of Earth: the panspermia hypothesis.

Recent Related Studies

In recent years, studies have been made of the amino acid composition of the products of "old" areas in "old" genes, defined as those that are found to be common to organisms from several widely separated species, assumed to share only the last universal ancestor (LUA) of all extant species. These studies found that the products of these areas are enriched in those amino acids that are also most readily produced in the Miller–Urey experiment. This suggests that the original genetic code was based on a smaller number of amino acids – only those available in prebiotic nature – than the current one.

Jeffrey Bada, himself Miller's student, inherited the original equipment from the experiment when Miller died in 2007. Based on sealed vials from the original experiment, scientists have been able to show that although successful, Miller was never able to find out, with the equipment available to him, the full extent of the experiment's success. Later researchers have been able to isolate even more different amino acids, 25 altogether. Bada has estimated that more accurate measurements could easily bring out 30 or 40 more amino acids in very low concentrations, but the researchers have since discontinued the testing. Miller's experiment was therefore a remarkable success at synthesizing complex organic molecules from simpler chemicals, considering that all life uses just 20 different amino acids.

In 2008, a group of scientists examined 11 vials left over from Miller's experiments of the early 1950s. In addition to the classic experiment, reminiscent of Charles Darwin's envisioned "warm little pond", Miller had also performed more experiments, including one with conditions similar to those of volcanic eruptions. This experiment had a nozzle spraying a jet of steam at the spark discharge. By using high-performance liquid chromatography and mass spectrometry, the group found more organic molecules than

Miller had. Interestingly, they found that the volcano-like experiment had produced the most organic molecules, 22 amino acids, 5 amines and many hydroxylated molecules, which could have been formed by hydroxyl radicals produced by the electrified steam. The group suggested that volcanic island systems became rich in organic molecules in this way, and that the presence of carbonyl sulfide there could have helped these molecules form peptides.

Amino Acids Identified

Below is a table of amino acids identified in the "classic" 1952 experiment as published by Miller in 1953, the 2008 re-analysis of vials from the volcanic spark discharge experiment, and the 2010 re-analysis of vials from the H_2S-rich spark discharge experiment.

	"Classic" 1952 experiment	Volcanic spark discharge experiment (2008)	H_2S-rich spark discharge experiment (2010)
Glycine	*	*	*
α-Alanine	*	*	*
β-Alanine	*	*	*
Aspartic acid	*	*	*
α-Aminobutyric acid	*	*	*
Serine		*	*
Isoserine		*	*
α-Aminoisobutyric acid		*	*
β-Aminoisobutyric acid		*	*
β-Aminobutyric acid		*	*
γ-Aminobutyric acid		*	*
Valine		*	*
Isovaline		*	*
Glutamic acid		*	*
Norvaline		*	
α-Aminoadipic acid		*	
Homoserine		*	
2-Methylserine		*	
β-Hydroxyaspartic acid		*	
Ornithine		*	
2-Methylglutamic acid		*	
Phenylalanine		*	
Homocysteic acid			*

S-methylcysteine			*
Methionine			*
Methionine sulfoxide			*
Methionine sulfone			*
Isoleucine			*
Leucine			*
Ethionine			*

Petasis Reaction

Petasis reaction	
Named after	Nicos A. Petasis
Reaction type	Coupling reaction
Identifiers	
Organic Chemistry Portal	petasis-reaction
RSC ontology ID	RXNO:0000232

The Petasis reaction (alternatively called the Petasis borono–Mannich (PBM) reaction) is the chemical reaction of an amine, aldehyde, and vinyl- or aryl-boronic acid to form substituted amines.

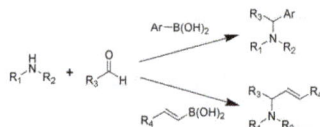

Reported in 1993 by Nicos Petasis as a practical method towards the synthesis of a geometrically pure antifungal agent, naftifine, the Petasis reaction can be described as a variation of the Mannich reaction. Rather than generating an enolate to form the substituted amine product, in the Petasis reaction, the vinyl group of the organoboronic acid serves as the nucleophile. In comparison to other methods of generating allyl amines, the Petasis reaction tolerates a multifunctional scaffold, with a variety of amines and organoboronic acids as potential starting materials. Additionally, the reaction does not require anhydrous or inert conditions. As a mild, selective synthesis, the Petasis reaction is useful in generating α-amino acids, and is utilized in combinatorial chemistry and drug discovery.

Reaction Mechanism

The mechanism of the Petasis reaction is not fully understood; however, it is similar to the Mannich reaction at its early stage. In the Mannich reaction, an imine or iminium salt serves as the electrophile to which the nucleophile adds; however, in the Petasis reaction it is not clear which intermediate serves as the electrophile. Petasis proposes that the reaction is characterized by a complex equilibrium among the three starting materials and various intermediates, and the final product is formed via a rate-determining and irreversible C-C bond formation step. The condensation between amine 1 and carbonyl 2 forms hemiaminal 4, which is in a complex equilibrium with iminium ion 3 and aminal 5. Boronic acid 6 react with hemiaminal 4 and aminal 5 in a reversible fashion via intermediate 7 and 8 respectively, forming again electrophilic iminium ion 3, this time accompanied by nucleophilic boronate 3'. Note that there are no evidences suggesting that boronic acid alone can directly react with iminium ions: In addition to needing acid for an appreciable amount of iminium salt to be generated, it has been shown that vinyl boronic acids do not react efficiently with preformed iminium salts (like Eschenmoser's salt). The irreversible C-C bond migration between 3 and 3'then follows, furnishing desired product 9 with loss of boric acid. All intermediates will ultimately lead to the final product, as the reaction between 3 and 3' is irreversible, pulling the equilibrium of the entire system towards the final product.

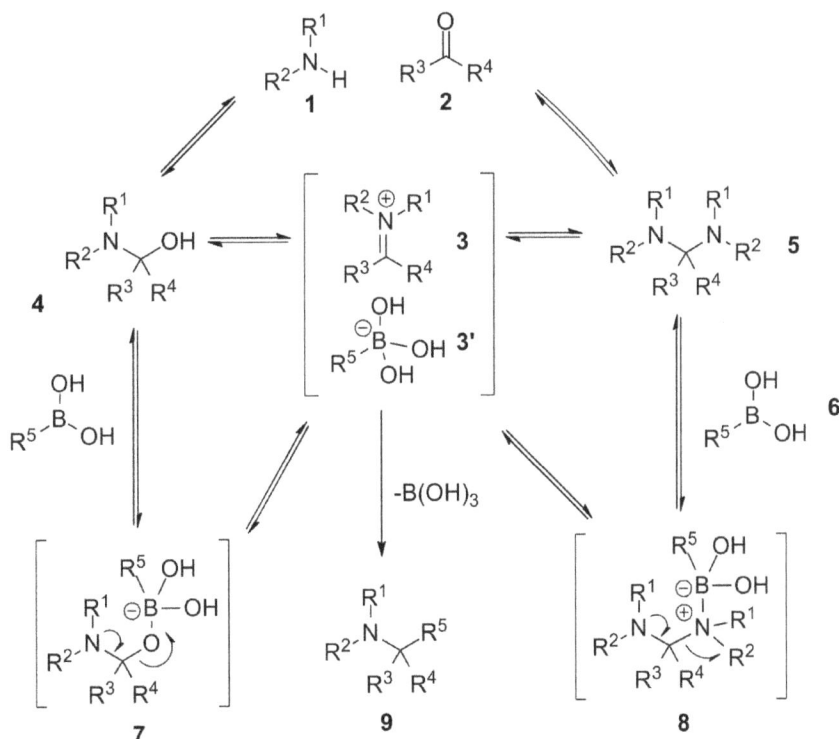

Density functional theory (DFT) studies have been performed to study the mechanism of Petasis reaction. Starting with the Petasis reaction between glyoxylic acid, dimethylamine and phenylboronic acid, Gois et al. reported that the migration of the boronic acid substituent (phenyl group) of the "ate complex" **A** incurs an energy barrier of 10kcal/mol and a five-membered transition state **B**. Formation of quaternary boron salts during the reaction has been experimentally confirmed by Hansen and coworkers. They reported that, in the absence of amine, an upfield ^{11}B shift is observed after the addition of glyoxylic acid to a solution of phenylboronic acid. This is presumably due to the formation of electron rich boronate species.

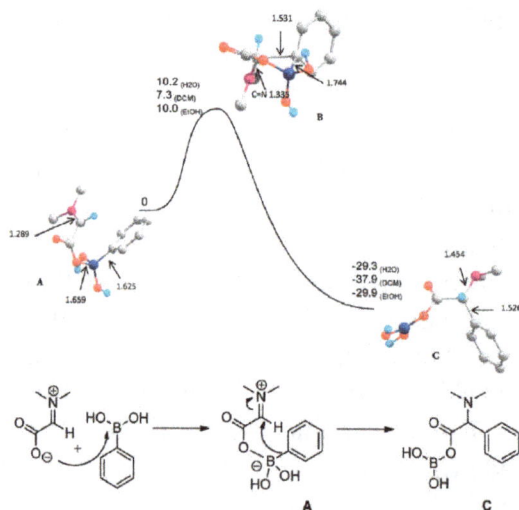

Preparation

The Petasis reaction proceeds under mild conditions, without the use of strong acids, bases, or metals. Unlike the Strecker and Ugi syntheses, the Petasis reaction avoids the use of cyanide and isocyanide reagents. The amine is mixed with the carbonyl substrate using either dioxane or toluene as a solvent at 90 °C for 10 minutes. Subsequently, the boronic acid is added to the mixture and product is generated, either after 30 minutes at 90 °C, or after several hours at 25 °C. In α-amino acid synthesis, α-keto acids, such as glyoxylic and pyruvic acid, are stirred in ethanol, toluene, or dichloromethane with amines and vinyl boronic acids at 25–50 °C for 12–48 h to give the corresponding β,γ-unsaturated compounds.

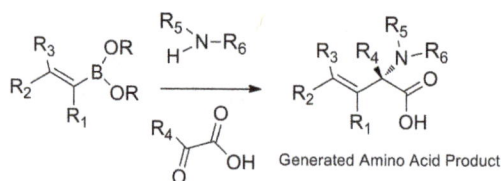

Generated Amino Acid Product

One of the most attractive features of the Petasis reaction is its use of boronic acids as a nucleophilic source. Unlike most vinyl substrates, vinyl boronic acids are stable to air and water and can be removed during workup with a simple extraction. Many boronic acid derivatives are easy to prepare and with the advent of the Suzuki coupling, a larger number of them are now commercially available. In the seminal report of the reaction, the organoboronic acids were prepared by hydroboration of terminal alkynes with catecholborane.

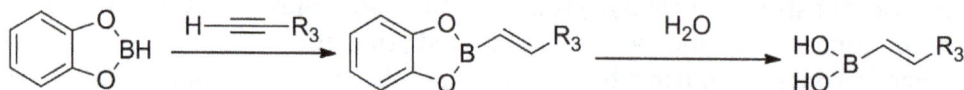

Other methods of generating boronic acids were also reported.

Reaction Scope and Synthetic Applications

A wide variety of functional groups including alcohols, carboxylic acids, and amines are tolerated in the Petasis Reaction. Known substrates that are compatible with reaction conditions include vinylboronate esters, arylboronate esters, and potassium organotrifluoroborates. Additionally, a variety of substituted amines can be used other than secondary amines. Tertiary aromatic amines, hydrazines, hydroxylamines, sulfonamides, and indoles have all been reported.

Synthesis of Allyl Amines

Petasis and coworkers proposed, in their seminal study, that vinyl boronic acids can react with the adducts of secondary amines and paraformaldehyde to give tertiary allylamines. The geometry of the double bond of the starting vinyl boronic acid is completely retained in the final product. Yield is typically in the good to excellent range. The following reaction is particularly effective, hitting a yield of 96%:

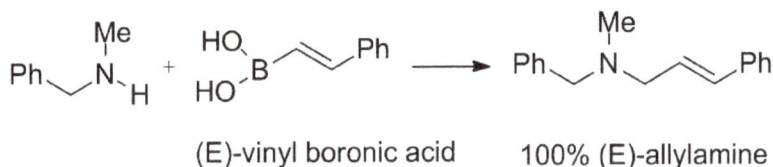

(E)-vinyl boronic acid   100% (E)-allylamine

Petasis and coworkers used this reaction to synthesize naftifine, a very potent topical antimycotic, in one step in 82% yield. Other compounds with related structure include terbinafine and NB598.

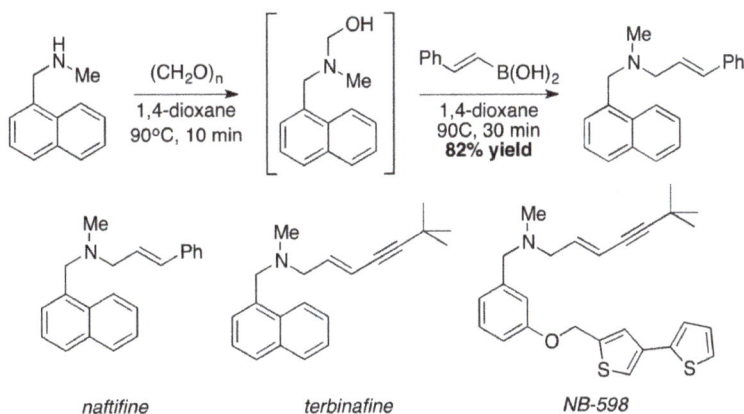

naftifine   terbinafine   NB-598

Synthesis of Amino Acids

β,γ-unsaturated, N-substituted amino acids are conveniently prepared through the condensation of organoboronic acids, boronates, or boronic esters with amines and glyoxylic acids. Yields are typically in the range of 60–80%, and a wide variety of polar or non-polar solvents can be employed (although DCM and MeOH is typically used). Free amino acids that do not have N-substitutions can be prepared by using trityl amine or bis-(4-methoxyphenyl)methyl amine, followed by deprotection under aciic conditions. Piettre and coworkers found out that the usage of highly polar protic solvents like Hexafluoroisopropanol (HFIP) can shorten reaction time and improve yield. Microwave irradiation was also be used to promote the reaction in methanol.

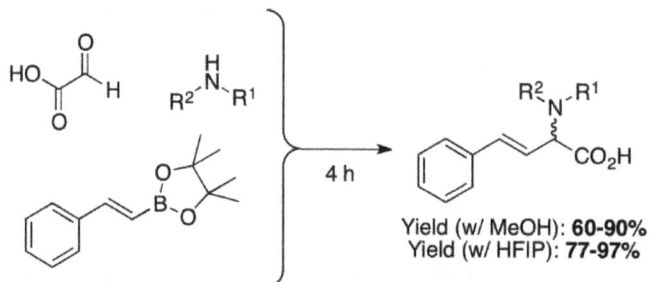

Yield (w/ MeOH): **60-90%**
Yield (w/ HFIP): **77-97%**

Apart from vinyl boronic acids, aryl boronic acids and other heterocyclic derivatives can also be used in Petasis multicomponent coupling. Possible substrate scope includes thienyl, pyridyl, furyl, and benzofuranyl, 1-naphthyl, and aryl groups with either electron-donating or electron-withdrawing substituent.

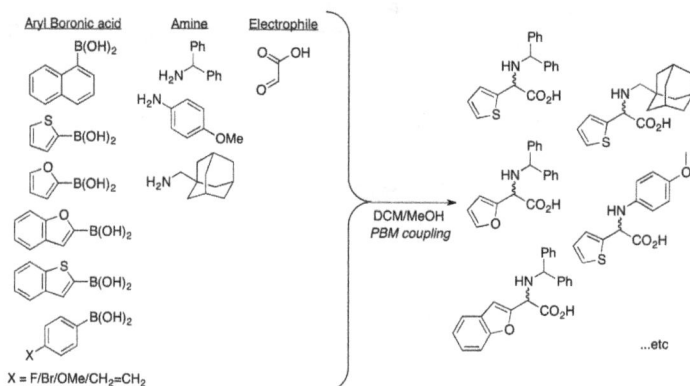

Clopidogrel, an antiplatelet agent, was racemically synthesized by Kalinski and coworkers in two steps, using Petasis reaction as the key strategy. Acid-catalyzed esterification immediately following the multicomponent coupling steps to afford Clopidogrel in 44% overall yield.

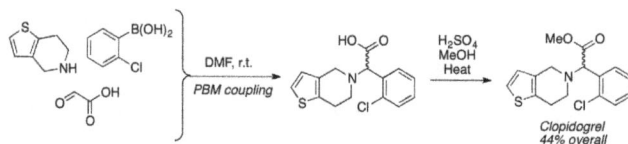

The Petasis reaction exhibits high degrees of stereocontrol when a chiral amine or aldehyde is used as a substrate. When certain chiral amines, such as (S)-2-phenylglycinol, are mixed with an α-keto acid and vinyl boronic acid at room temperature, the corresponding allylamine is formed as a single diastereomer. Furthermore, enantiomeric purity can be achieved by hydrogenation of the diastereoselective product. In the reaction with (S)-2-phenylglycinol, (R)-2-phenylglycinol is generated in 76% yield.

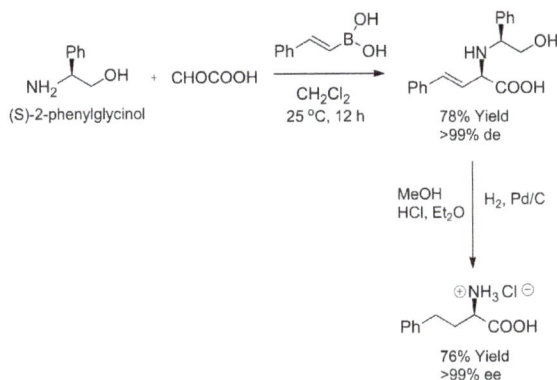

Unconventional Synthesis of Carboxylic Acids

Apart from amino-acids, Petasis borono-Mannich reaction can also be used to prepare carboxylic acids, albeit with unconventional mechanisms. Naskar et al. reported the use of *N*-substituted indoles as amine equivalent. The mechanism begins with the nucleophilic attack of the 3-position of the "N"-substituted indole to electrophilic aldehyde, followed by formation of "ate complex" 1 via the reaction of boronic acid with the carboxylic acid. The intermediate then undergoes dehydration, followed by migration of boronate-alkyl group to furnish the final carboxylic acid product. The yield is in the moderate to good range (40–70%). A wide range of aryl boronic acids is tolerated, while the usage of vinyl boronic acids is not reported. It is interesting to note that "N"-unsubstituted indoles react very sluggishly under normal reaction conditions, thus confirming the mechanism below.

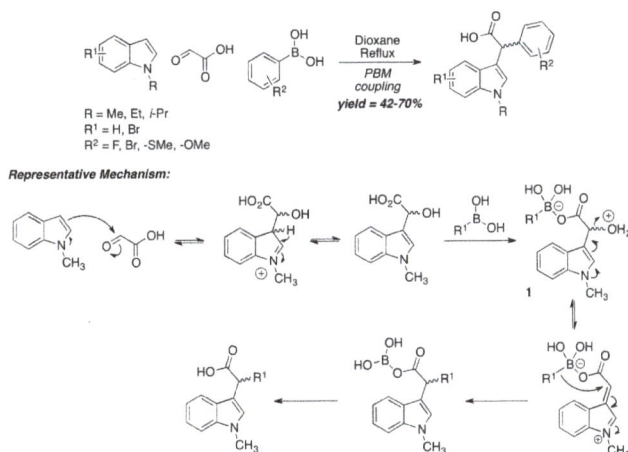

Naskar et al. also proposed the usage of tertiary aromatic amines in Petasis reaction as another equivalent of amine nucleophile. The mechanism is similar to the N-substituted indole case. The reaction is carried out under harsh conditions (24-hr reflux in 1,4-dioxane), but the resultant carboxylic acid is obtained in reasonable yield (41–54% yield). Note that the usage of α-ketoacids instead of glyoxylic acid does not diminish yields. 1,3,5-trioxygenated benzene derivatives can also be used in lieu of tertiary aromatic amines.

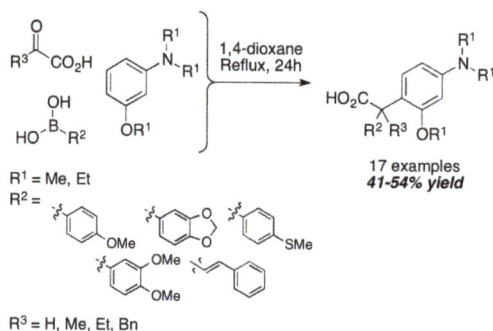

R^1 = Me, Et

R^2 =

R^3 = H, Me, Et, Bn

Synthesis of Iminodicarboxylic Acid Derivatives

When used as nitrogen nucleophiles, amino acids can furnish various iminodicarboxylic acid derivatives. High diastereoselectivity is usually observed, and the newly formed stereocenter usually share the same configuration with the starting amino acid. This reaction works well in highly polar solvents (ex. water, ethanol, etc.). Peptides with unprotected nitrogen terminal can also be used as a nitrogen nucleophile equivalent. Petasis and coworkers prepared Enalaprilat, an ACE inhibitor, with this method.

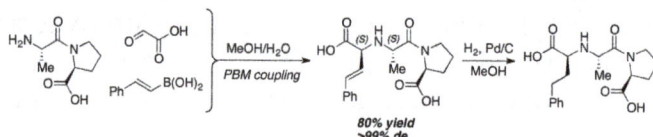

Synthesis of Peptidomimetic Heterocycles

When diamines are used in PBM reactions, heterocycles of various structures, such as piperazinones, benzopiperazinones, and benzodiazepinones, are efficiently prepared. Lactamization reactions are commonly employed to form the heterocycles, usually under strongly acidic conditions.

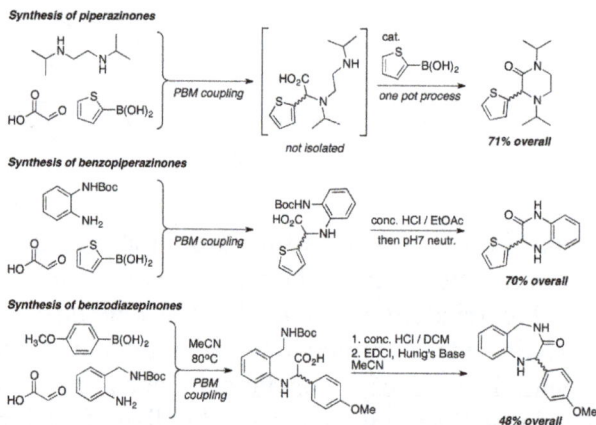

Synthesis of Amino Alcohols

When a α-hydroxy aldehyde is used as a substrate in the synthesis of β-amino alcohols, a single diastereomer is generated. This reaction forms exclusively anti-product, confirmed by ¹H NMR spectroscopy. The product does not undergo racemization, and when enantiomerically pure α-hydroxy aldehydes are used, enantiomeric excess can be achieved. It is believed that the boronic acid first reacted with the chiral hydroxyl group, furnishing a nucleophilic alkenyl boronate, followed by face selective, intramolecular migration of the alkenyl group into the electrophilic iminium carbon, forming the desired C-C bond irreversibly. In the reaction of enantiomerically pure glyceraldehydes, the corresponding 3-amino 1,2-diol product is formed in 70% yield and greater than 99% ee.

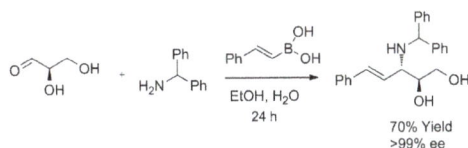

Pyne and coworkers suggested that diastereoselectivity arises from the reaction of the more stable (and, in this case, more reactive) conformation of the ate complex, where 1,3 allylic strain is minimized.

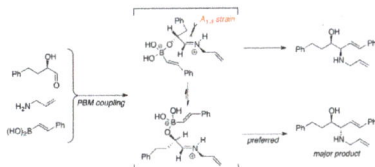

Using dihydroxyacetone, a somewhat unconventional aldehyde equivalent, Sugiyama et al. is able to use Petasis reaction to assemble the core structure of FTY720 (a potent immunosuppressive agent) in 40% yield. A straightforward hydrogenation then follows to afford the product via a one-step benzyl-group removal and C-C double bond hydrogenation.

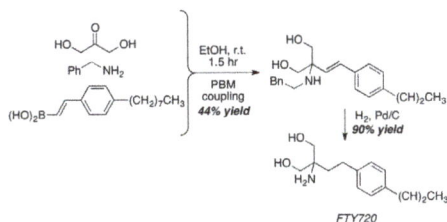

Synthesis of Amino Polyols and Amino Sugars

Petasis and coworkers reported the usage of unprotected carbohydrates as the carbonyl component in PBM reactions. It is used as the equivalent of α-hydroxyl aldehydes with pre-existing chirality, and the aminopolyol product is usually furnished with moderate to good yield, with excellent selectivity. A wide variety of carbohydrates, as well as nitrogen nucleophiles (ex. amino acids), can be used to furnish highly stereochemically-enriched products. The aminopolyol products can then undergo further reactions to prepare aminosugars. Petasis used this reaction to prepare Boc-protected mannosamine from D-arabinose.

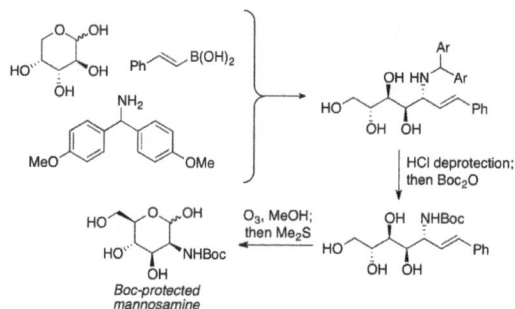

Applications in Enantioselective Synthesis

With chiral amine nucleophile

Generally speaking, when chiral amine is used in Petasis coupling, the stereochemical outcome of Petasis reaction is strongly correlated to the chirality of the amine, and high to excellent diastereoselectivity is observed even without the usage of bulky chiral inducing groups. Chiral benzyl amines, 2-substituted pyrrolidines, and 5-substituted 2-morpholinones have been shown to induce good to excellent diastereomeric excess under different Petasis reaction conditions.

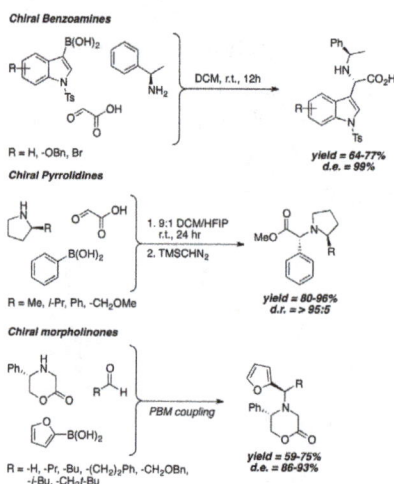

With Chiral N-acyliminium Ions

Chiral N-acyliminium ion "starting materials" are generally prepared by in-situ dehydration of cyclic hemiaminal. They also carry a chiral hydroxyl group that is in proximity with the iminium carbon; boronic acids react with such chiral hydroxyl groups to form a chiral and electron-rich boronate species, followed by side-selective and intramolecular boronate vinyl/aryl transfer into the iminium carbon. Hence, the reaction is highly diastereoselective, with cis- boronate aryl/vinyl transfer being the predominant pathway. Hydroxypyrrolidines and Hydroxy-γ- and δ-lactams have been shown to react very diastereoselectively, with good to excellent yield. However, such procedures are limited to the usage of vinyl- or electron-rich aryl- boronic acids only.

Hydroxy δ-lactam as substrates:

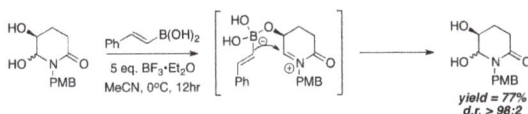

Hydroxy pyrrolidines as substrates:

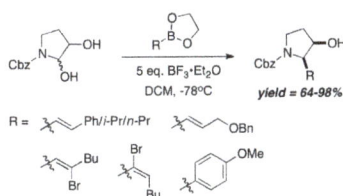

Batey and coworkers take advantage of the high diastereoselectivity of this reaction to prepare (±)-6-deoxycastanospermine in 7 steps, with an impressive overall yield of 32% (from the vinyl boronic ester). The key acyclic precursor to deoxycastanospermine (A) is formed first by condensing vinyl boronic ester 1 with Cbz-protected hydroxy-pyrrolidine 2 with a PBM coupling, followed by dihydroxylation and TBS protetction. A then undergo intramolecular cyclization via a one-pot imine formation and reduction sequel, followed by TBS deprotection, to afford (±)-6-deoxycastanospermine.

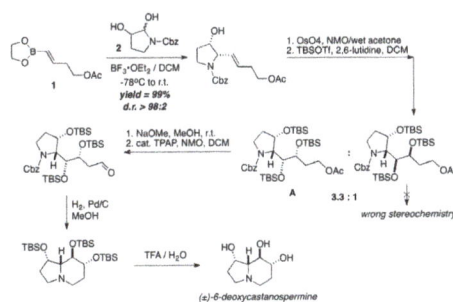

With Thiourea Catalyst

Takemoto and coworkers of Kyoto University recently reported an enantioselective Petasis-type reaction to transform quinolines into respective chiral 1,2-dihydroquinolines (product) using alkenyl boronic acids and chiral thiourea catalyst. Good yields (59–78%) and excellent enantioselectivities (82–96%) are reported.

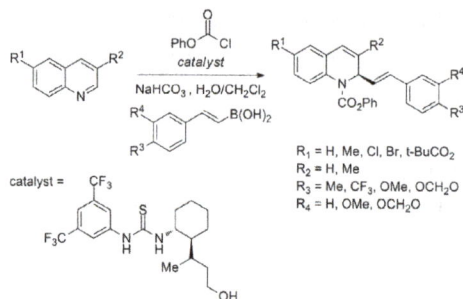

Takemoto and coworkers observed that addition of chloroformates are required as electrophilic activating agents, and the reaction does not proceed without them. Also, a 1,2-amino alcohol functionality is required on the catalyst for the reaction to proceed stereoselectively. They rationalize these findings by suggesting that the chloroformate reagent reacted with the quinoline nitrogen to make an N-acyated quinolinium intermediate B, which is further activated by electrophilic chiral thiourea. They also suggest that the 1,2-aminoalcohol functionality of the catalyst is chelating to the alkenyl boronic acids and that such chelation directed the stereochemical outcome.

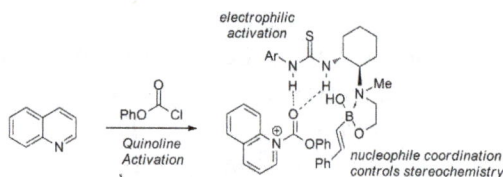

With Chiral Biphenols

Schaus and Lou of Boston University reported the following reaction, in which chiral α-amino acids with various functionalities are conveniently furnished by mixing alkenyl diethyl boronates, secondary amines, glyoxylates and chiral biphenol catalyst in toluene in one-pot:

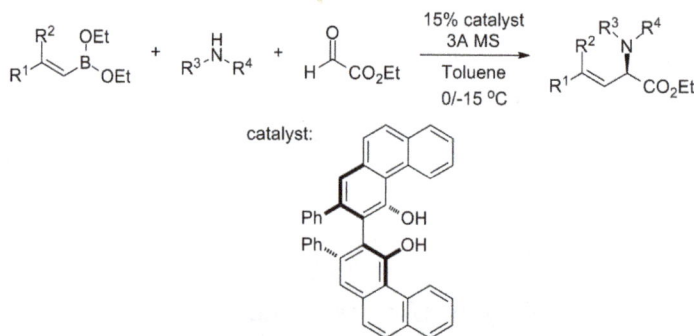

This reaction tolerates a wide range of functionalities, both on the sides of alkenyl boronates and the secondary amine: the electron-richness of the substrates does not affect

the yield and enantioselectivity, and sterically demanding substrates (dialkylsubstitut-
ed alkenyl boronates and amines with α-stereocenter) only compromise enantioselectivity slightly. Reaction rates do vary on a case-by-case basis.

Interestingly, under the reported condition, boronic acids substrates failed to give any
enantioselectivity. Also, 3Å molecular sieve is used in the reaction system. While the
authors did not provide the reason for such usage in the paper, it was speculated that
3Å molecular sieves act as water scavenger and prevent the decomposition of alkenyl
diethyl boronates into their respective boronic acids. The catalyst could be recycled
from the reaction and reused without compromising yield or enantioselectivity.

More recently, Yuan with coworkers from Chengdu Institute of Organic Chemistry,
Chinese Academy of Science combined both approaches (chiral thiourea catalyst and
chiral biphenol) in a single catalyst, reporting for the first time the catalytic system that
is capable of performing enantioselective Petasis reaction between salicylaldehydes,
cyclic secondary amines and aryl- or alkenylboronic acids:

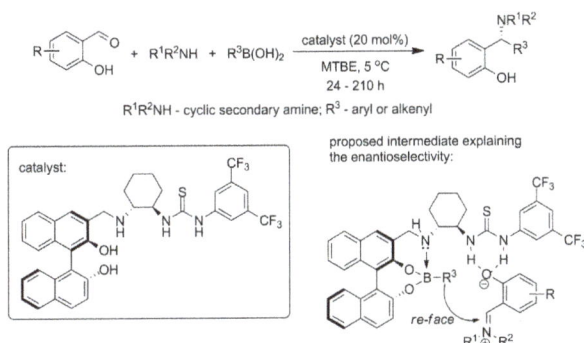

In one application the Petasis reaction is used for quick access to a multifunctional
scaffold for divergent synthesis. The reactants are the lactol derived from L-phenyl-
lactic acid and acetone, l-phenylalanine methyl ester and a boronic acid. The reaction
takes place in ethanol at room temperature to give the product, an anti-1,2-amino
alcohol with a diastereomeric excess of 99%.

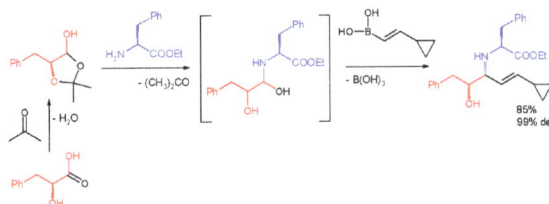

Notice that the authors cannot assess syn-1,2-amino alcohol with this method due to
intrinsic mechanistic selectivity, and the authors argue that such intrinsic selectivity
hampers their ability to access the full matrix of stereoisomeric products for the
usage of small molecule screening. In a recent report, Schaus and co-workers
reported that syn amino alcohol can be obtained with the following reaction
condition, using a chiral dibromo-biphenol catalyst their group developed:

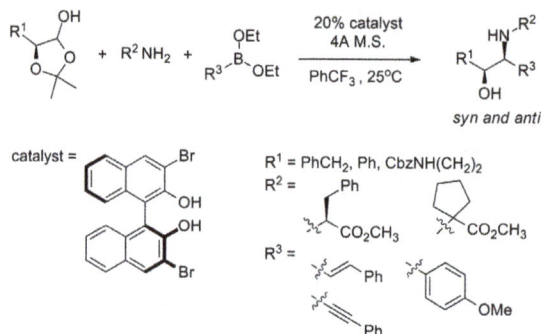

Although the syn vs. anti diastereomeric ratio ranges from mediocre to good (1.5:1 to 7.5:1), the substrate scope for such reactions remain rather limited, and the diastereoselectivity is found to be dependent on the stereogenic center on the amine starting material.

Petasis Reaction and Total Synthesis

Beau and coworkers assembled the core dihydropyran framework of zanamivir congeners via a combination of PBM reaction and Iron(III)-promoted deprotection-cyclization sequence. A stereochemically-defined α-hydroxyaldehyde 2, diallylamine and a dimethylketal-protected boronic acid 1 is coupled to form the acyclic, stereochemically-defined amino-alcohol 3, which then undergoes an Iron(III)-promoted cyclization to form a bicyclic dihydropyran 4. Selective opening of the oxazoline portion of the dihydropyran intermediate 4 with water or timethylsilyl azide then furnish downstream products that have structures resembling the Zanamivir family members.

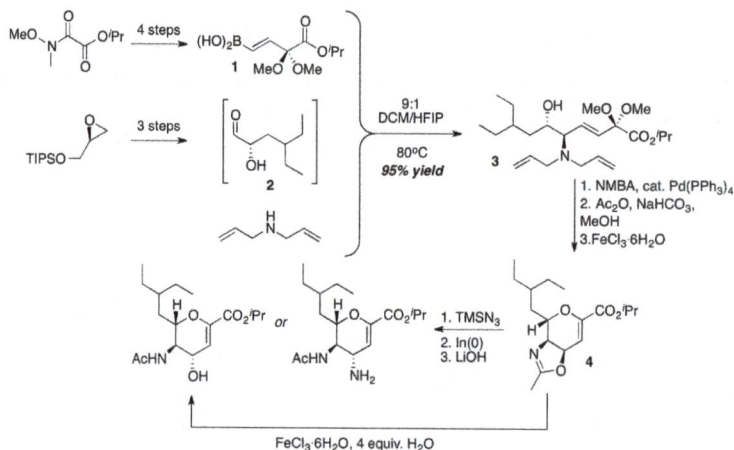

Wong and coworkers prepared N-acetylneuraminic acid with a PBM coupling, followed by nitrone-[3+2] cycloaddition. Vinylboronic acid is first coupled with L-arabinose 1 and Bis(4-methoxyphenyl) methanamine 2 to form an stereochemically-defined allyl amine 3. Afterwards, the sequence of dipolar cycloaddition, base-mediated N-O bond breakage and hydrolysis then complete the synthesis of N-acetylneuraminic acid.

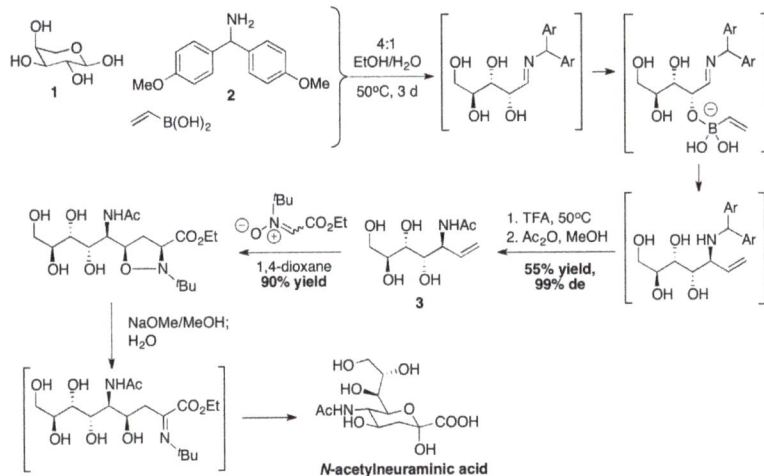

Schöllkopf Method

The Schöllkopf method or Schöllkopf Bis-Lactim Amino Acid Synthesis is a method in organic chemistry for the asymmetric synthesis of chiral amino acids. The method was established in 1981 by Ulrich Schöllkopf. In it glycine is a substrate, valine a chiral auxiliary and the reaction taking place an alkylation.

Reaction Mechanism

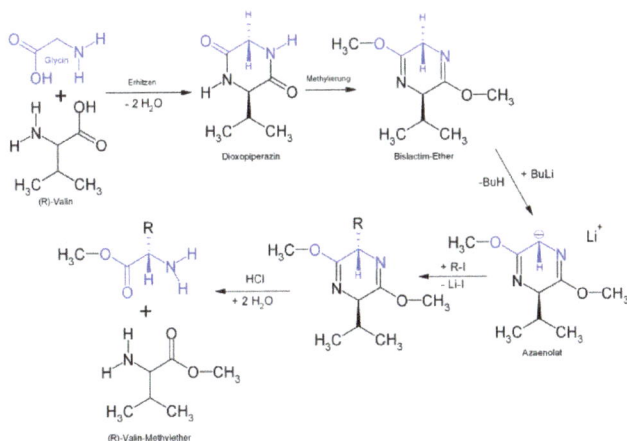

Reaction mechanism for the Schöllkopf method

The dipeptide derived from glycine and (R-)valine is converted into a 2,5-Diketopiperazine (a cyclic dipeptide). Double O-methylation gives the bis-lactim. A proton is then abstracted from the prochiral position on glycine with *n*-BuLi. The next step decides the stereoselectivity of the method: One face of the carbanionic center is shielded by steric hindrance from the isopropyl residue on valinine. The reaction of the anion with an alkyl iodide will form the alkylated product with a strong preference for just one enantiomer. In the final step the dipeptide is cleaved by acidic hydrolysis in two amino acid methyl esters which can be separated from each other.

With valine Schöllkopf selected the natural proteinogenic amino acid with the largest non-reactive and nonchiral residue in order to achieve the largest possible stereoselectivity, generally speaking enantiomeric excess of over 95% ee is feasible.

With the Schöllkopf method all amino acids can be synthesised when a suitable R-I reagent is available. R does not need to be an alkyl group but can also be more complicated. The method is limited to the laboratory for the synthesis of exotic amino acids. Industrial applications are not known. One disadvantage is limited atom economy.

Strecker Amino Acid Synthesis

Strecker synthesis	
Named after	Adolph Strecker
Reaction type	Substitution reaction
Identifiers	
Organic Chemistry Portal	strecker-synthesis
RSC ontology ID	RXNO:0000207

The Strecker amino acid synthesis, also known simply as the Strecker synthesis, was devised by German chemist Adolph Strecker, and is a term used for a series of chemical reactions that synthesize an amino acid from an aldehyde or ketone. The aldehyde is condensed with ammonium chloride in the presence of potassium cyanide to form an α-aminonitrile, which is subsequently hydrolyzed to give the desired amino acid. In the original Strecker reaction acetaldehyde, ammonia, and hydrogen cyanide combined to form after hydrolysis alanine.

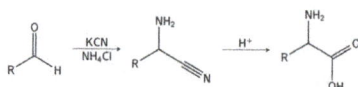

While usage of ammonium salts gives unsubstituted amino acids, primary and secondary amines also successfully give substituted amino acids. Likewise, the usage of ketones, instead of aldehydes, gives α,α-disubstituted amino acids.

The traditional synthesis of Adolph Strecker from 1850 gives racemic α-amino nitriles, but several procedures utilizing asymmetric auxiliaries or asymmetric catalysts have been developed.

Reaction Mechanism

In the first part of the reaction, the carbonyl oxygen of an aldehyde is protonated, followed by a nucleophilic attack of ammonia to the carbonyl carbon. After subsequent proton exchange, water is cleaved from the iminium ion intermediate. A cyanide ion then attacks the iminium carbon yielding an aminonitrile.

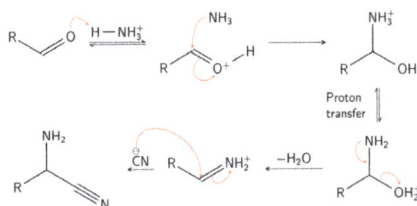

In the second part of the Strecker Synthesis the nitrile nitrogen of the aminonitrile is protonated, and the nitrile carbon is attacked by a water molecule. A 1,2-diamino-diol is then formed after proton exchange and a nucleophilic attack of water to the former nitrile carbon. Ammonia is subsequently eliminated after the protonation of the amino group, and finally the deprotonation of a hydroxyl group produces an amino acid.

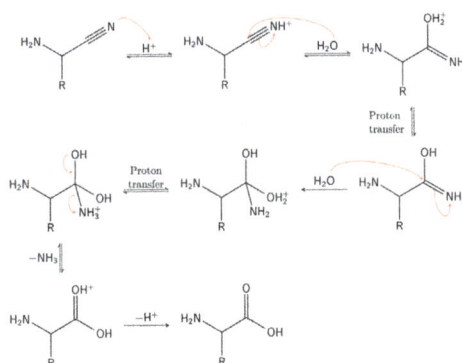

One example of the Strecker synthesis is a multikilogram scale synthesis of an L-valine derivative starting from *3-methyl-2-butanone*:

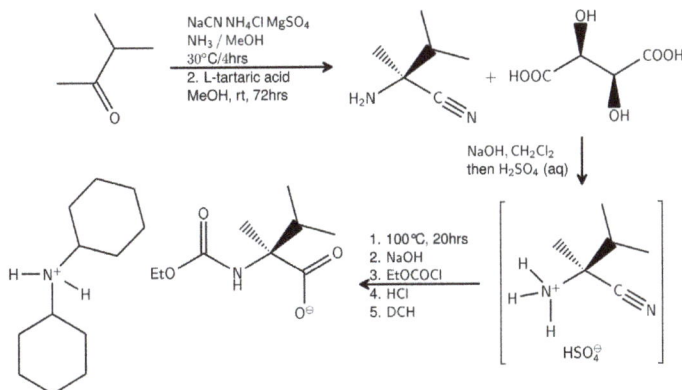

Asymmetric Strecker Reactions

The asymmetric Strecker reaction was pioneered by Kaoru Harada in 1963. By replacing ammonia with (S)-alpha-phenylethylamine as chiral auxiliary the ultimate reaction product was chiral alanine. The first asymmetric synthesis via a chiral catalyst was reported in 1996.

Catalytic Asymmetric Strecker Reactions

Catalytic asymmetric Strecker reaction can be effected using thiourea-derived catalyst. In 2012, a BINOL-derived catalyst was employed to generate chiral cyanide anion.

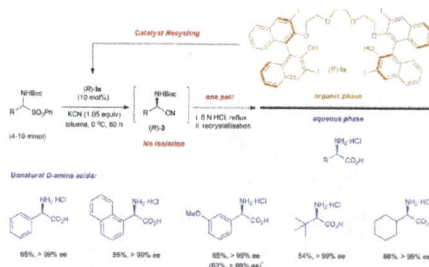

Catalytic Asymmetric Strecker Synthesis-Nature Chem

Enantioselective Synthesis

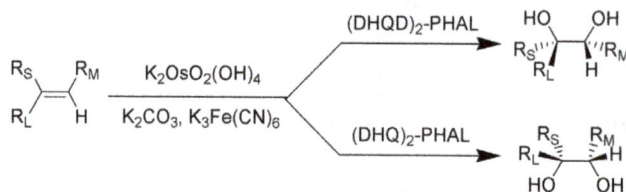

In the Sharpless dihydroxylation reaction the chirality of the product can be controlled by the "AD-mix" used. This is an example of enantioselective synthesis using asymmetric induction Key: R_L = Largest substituent; R_M = Medium-sized substituent; R_S = Smallest substituent

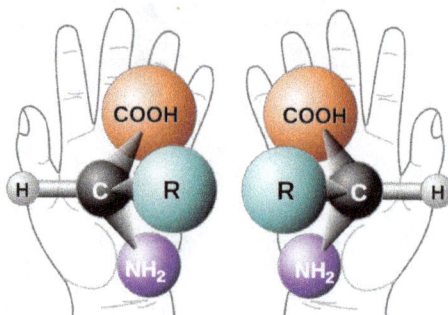

Two enantiomers of a generic alpha amino acid

Enantioselective synthesis, also called chiral synthesis or asymmetric synthesis, is a form of chemical synthesis. It is defined by IUPAC as: a chemical reaction (or reaction sequence) in which one or more new elements of chirality are formed in a substrate molecule and which produces the stereoisomeric (enantiomeric or diastereoisomeric) products in unequal amounts.

Put more simply: it is the synthesis of a compound by a method that favors the formation of a specific enantiomer or diastereomer.

Enantioselective synthesis is a key process in modern chemistry and is particularly important in the field of pharmaceuticals, as the different enantiomers or diastereomers of a molecule often have different biological activity.

Overview

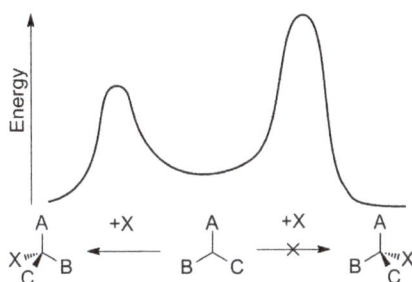

A Gibbs free energy plot of an enantioselective addition reaction.

Many of the building blocks of biological systems, such as sugars and amino acids, are produced exclusively as one enantiomer. As a result of this living systems possess a high degree of chemical chirality and will often react differently with the various enantiomers of a given compound. Examples of this selectivity include:

- Flavour: the artificial sweetener aspartame has two enantiomers. L-aspartame tastes sweet, yet D-aspartame is tasteless

- Odor: R-(−)-carvone smells like spearmint yet S-(+)-carvone, smells like caraway.

- Drug effectiveness: the antidepressant drug Citalopram is sold as a racemic mixture. However, studies have shown that only the (S)-(+) enantiomer is responsible for the drug's beneficial effects.

- Drug safety: Dpenicillamine is used in chelation therapy and for the treatment of rheumatoid arthritis. However Lpenicillamine is toxic as it inhibits the action of pyridoxine, an essential B vitamin.

As such enantioselective synthesis is of great importance; but it can also be difficult to achieve. Enantiomers possess identical enthalpies and entropies, and hence should be produced in equal amounts by an undirected process – leading to a racemic mixture.

The solution is to introduce a chiral feature which will promote the formation of one enantiomer over another via interactions at the transition state. This is known as asymmetric induction and can involve chiral features in the substrate, reagent, catalyst or environment and works by making the activation energy required to form one enantiomer lower than that of the opposing enantiomer.

Asymmetric induction can occur intramolecularly when given a chiral starting material. This behaviour can be exploited, especially when the goal is to make several consecutive chiral centres to give a specific enantiomer of a specific diastereomer. An aldol reaction, for example, is inherently diastereoselective; if the aldehyde is enantiopure, the resulting aldol adduct is diastereomerically and enantiomerically pure.

Approaches

Enantioselective Catalysis

In general, enantioselective catalysis (known traditionally as asymmetric catalysis) refers to the use of chiral coordination complexes as catalysts. This approach is very commonly encountered, as it is effective for a broader range of transformations than any other method of enantioselective synthesis. The catalysts are typically rendered chiral by using chiral ligands, however it is also possible to generate chiral-at-metal complexes using simpler achiral ligands. Most enantioselective catalysts are effective at low concentrations making them well suited to industrial scale synthesis; as even exotic and expensive catalysts can be used affordably. Perhaps the most versatile example of enantioselective synthesis is asymmetric hydrogenation, which is able to reduce a wide variety of functional groups.

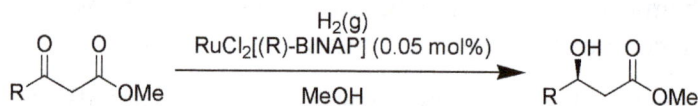

With only 75 natural metals in existence (and not all of these showing extensive catalytic activities) the design of new catalysts is very much dominated by the development of new classes of ligands. Certain ligands, often referred to as 'privileged ligands', have been found to be effective in a wide range of reactions; examples include BINOL, Salen and BOX. However, most catalysts are rarely general, requiring certain functional groups in the substrate to form the transition state complex correctly: arbitrary structures cannot be used. For example, Noyori asymmetric hydrogenation with BINAP/Ru requires a β-ketone, although another catalyst, BINAP/diamine-Ru, widens the scope to α,β-olefins and aromatics.

Chiral Auxiliaries

A chiral auxiliary is an organic compound which couples to the starting material to form new compound which can then undergo enantioselective reactions via intramolecular asymmetric induction. At the end of the reaction the auxiliary is removed, under conditions that will not cause racemization of the product. It is typically then recovered for future use.

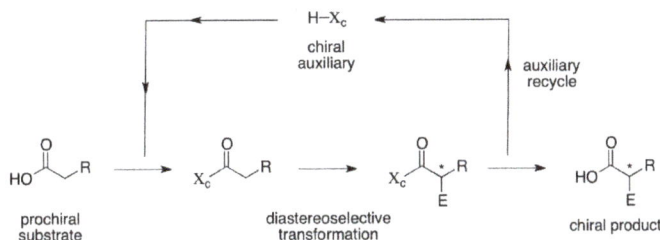

Chiral auxiliaries must be used in stoichiometric amounts to be effective and require additional synthetic steps to append and remove the auxiliary. However, in some cases the only available stereoselective methodology relies on chiral auxiliaries and these reactions tend to be versatile and very well-studied, allowing the most time-efficient access to enantiomerically pure products. Additionally, the products of auxiliary-directed reactions are diastereomers, which enables their facile separation by methods such as column chromatography or crystallization.

Biocatalysis

Biocatalysis makes use of biological compounds, ranging from isolated enzymes to living cells, to perform chemical transformations. The advantages of these reagents include very high ee's and reagent specificity, as well as mild operating conditions and low environmental impact. Biocatalysts are more commonly used in industry than in academic research; for example in the production of statins. The high reagent specificity can be a problem however; as it often requires that a wide range of biocatalysts be screened before an effective reagent is found.

Enantioselective Organocatalysis

Organocatalysis refers to a form of catalysis, where the rate of a chemical reaction is increased by an organic compound consisting of carbon, hydrogen, sulfur and other non-metal elements. When the organocatalyst is chiral enantioselective synthesis can be achieved; for example a number of carbon–carbon bond forming reactions become enantioselective in the presence of proline with the aldol reaction being a prime example. Organocatalysis often employs natural compounds and secondary amines as chiral catalysts; these are inexpensive and environmentally friendly, as no metals are involved.

Chiral Pool Synthesis

Chiral pool synthesis is one of the simplest and oldest approaches for enantioselective synthesis. A readily available chiral starting material is manipulated through successive reactions, often using achiral reagents, to obtain the desired target molecule. This can meet the criteria for enantioselective synthesis when a new chiral species is created, such as in an S_N2 reaction.

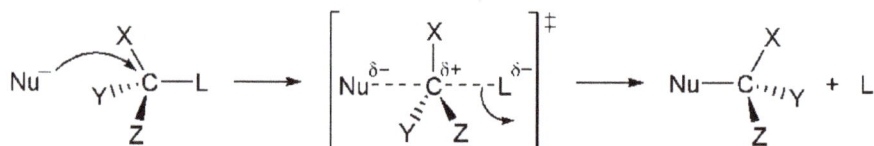

Chiral pool synthesis is especially attractive for target molecules having similar chirality to a relatively inexpensive naturally occurring building-block such as a sugar or amino acid. However, the number of possible reactions the molecule can undergo is restricted and tortuous synthetic routes may be required (e.g. Oseltamivir total synthesis). This approach also requires a stoichiometric amount of the enantiopure starting material, which can be expensive if it is not naturally occurring.

Alternative Approaches

Alternatives to enantioselective synthesis usually involve the isolation of one enantiomer from a racemic mixture by any of a number of methods. If the cost in time and money of making such racemic mixtures is low (or if both enantiomers may find use) then this approach may remain cost-effective. Common methods of separation are based around chiral resolution or kinetic resolution.

Separation and Analysis of Enantiomers

The two enantiomers of a molecule possess the same physical properties (e.g. melting point, boiling point, polarity etc.) and so behave identically to each other. As a result, they will migrate with an identical R_f in thin layer chromatography and have identical retention times in HPLC and GC. Their NMR and IR spectra are identical.

This can make it very difficult to determine whether a process has produced a single enantiomer (and crucially which enantiomer it is) as well as making it hard to separate enantiomers from a reaction which has not been 100% enantioselective. Fortunately, enantiomers behave differently in the presence of other chiral materials and this can be exploited to allow their separation and analysis.

Enantiomers do not migrate identically on chiral chromatographic media, such as quartz or standard media that has been chirally modified. This forms the basis of chiral column chromatography, which can be used on a small scale to allow analysis via GC and HPLC, or on a large scale to separate chirally impure materials. However this process can require large amount of chiral packing material which can be expensive. A common alternative is to use a chiral derivatizing agent to convert the enantiomers into a diastereomers, in much the same way as chiral auxiliaries. These have different physical properties and hence can be separated and analysed using conventional methods. Special chiral derivitizing agents known as 'chiral resolution agents' are used in the NMR spectroscopy of stereoisomers, these typically involve coordination to chiral europium complexes such as $Eu(fod)_3$ and $Eu(hfc)_3$.

The enantiomeric excess of a substance can also be determined using certain optical methods. The oldest method for doing this is to use a polarimeter to compare the level of optical rotation in the product against a 'standard' of known composition. It is also possible to perform ultraviolet-visible spectroscopy of stereoisomers by exploiting the Cotton effect.

One of the most accurate ways of determining the chirality of compound is to determine its absolute configuration by Xray Crystallography. However this is a labour-intensive process which requires that a suitable single crystal be grown.

History

Inception (1815–1905)

In 1815 the French physicist Jean-Baptiste Biot showed that certain chemicals could rotate the plane of a beam of polarised light, a property called optical activity. The nature of this property remained a mystery until 1848, when Louis Pasteur proposed that it had a molecular basis originating from some form of *dissymmetry*, with the term *chirality* being coined by Lord Kelvin a year later. The origin of chirality itself was finally described in 1874, when Jacobus Henricus van 't Hoff and Joseph Le Bel independently proposed the tetrahedral geometry of carbon; structural models prior to this work had been two-dimensional, and van 't Hoff Le Bel theorized that the arrangement of groups around this tetrahedron could dictate the optical activity of the resulting compound.

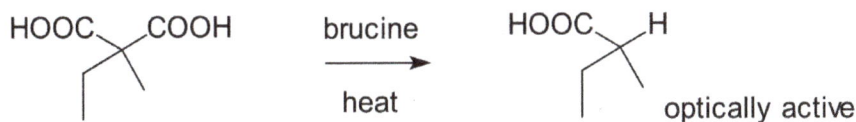

Marckwald's brucine-catalyzed enantioselective decarboxylation of 2-ethyl-2-methylmalonic acid, resulting in a slight excess of the levorotary form of the 2-methylbutyric acid product.

In 1894 Hermann Emil Fischer outlined the concept of asymmetric induction; in which he correctly ascribed selective the formation of D-glucose by plants to be due to the influence of optically active substances within chlorophyll. Fischer also successfully performed what would now be regarded as the first example of enantioselective synthesis, by enantioselectively elongating sugars via a process which would eventually become the Kiliani–Fischer synthesis.

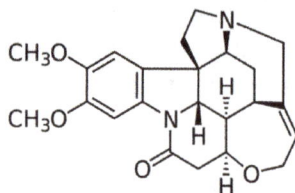

Brucine, an alkaloid natural product related to strychnine, used successfully as an organocatalyst by Marckwald in 1904.

The first enantioselective chemical synthesis is most often attributed to Willy Marckwald, Universität zu Berlin, for a brucine-catalyzed enantioselective decarboxylation of *2-ethyl-2-methylmalonic acid* reported in 1904. A slight excess of the levorotary form of the product of the reaction, 2-methylbutyric acid, was produced; as this product is also a natural product—e.g., as a side chain of lovastatin formed by its diketide synthase (LovF) during its biosynthesis—this result constitutes the first recorded total synthesis with enantioselectivity, as well other firsts (as Koskinen notes, first "example of asymmetric catalysis, enantiotopic selection, and organocatalysis"). This observation is also of historical significance, as at the time enantioselective synthesis could only be understood in terms of vitalism. Natural and artificial compounds were fundamentally different, it was argued, and chirality could only exist in natural compounds. Unlike Fischer, Marckwald had performed an enantioselective reaction upon an achiral, *un-natural* starting material, albeit with a chiral organocatalyst (as we now understand this chemistry).

Early Work (1905–1965)

The development of enantioselective synthesis was initially slow, largely due to the limited range of techniques available for their separation and analysis. Diastereomers possess different physical properties, allowing separation by conventional means, however at the time enantiomers could only be separated by spontaneous resolution (where enantiomers separate upon crystallisation) or kinetic resolution (where one enantiomer is selectively destroyed). The only tool for analysing enantiomers was optical activity using a polarimeter, a method which provides no structural data.

It was not until the 1950s that major progress really began. Driven in part by chemists such as R. B. Woodward and Vladimir Prelog but also by the development of new techniques. The first of these was Xray Crystallography, which was used to determine the absolute configuration of an organic compound by Johannes Bijvoet in 1951. Chiral chromatography was introduced a year later by Dalgliesh, who used paper chromatography to separate chiral amino acids. Although Dalgliesh was not the first to observe such separations, he correctly attributed the separation of enantiomers to differential retention by the chiral cellulose. This was expanded upon in 1960, when Klem and Reed first reported the use of chirally-modified silica gel for chiral HPLC chromatographic separation.

Thalido Mide

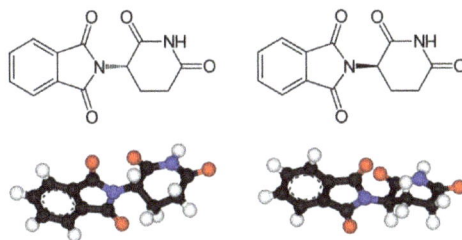

The two enantiomers of thalidomide:Left: (*S*)-thalidomide
Right: (*R*)-thalidomide

While it had long been known that the different enantiomers of a drug could have different activities, this was not accounted for in early drug design and testing. However following the thalidomide disaster the development and licensing of drugs changed dramatically.

First synthesized in 1953, thalidomide was widely prescribed for morning sickness from 1957 to 1962, but was soon found to be seriously teratogenic, eventually causing birth defects in more than 10,000 babies. The disaster prompted many counties to introduce tougher rules for the testing and licensing of drugs, such as the Kefauver-Harris Amendment (U.S.) and Directive 65/65/EEC1 (E.U.).

Early research into the teratogenic mechanism, using mice, suggested that one enantiomer of thalidomide was teratogenic while the other possessed all the therapeutic activity. This theory was later shown to be incorrect and has now been superseded by a body of research. However it raised the importance of chirality in drug design, leading to increased research into enantioselective synthesis.

Modern Age (Since 1965)

The Cahn−Ingold−Prelog priority rules (often abbreviated as the CIP system) were first published in 1966; allowing enantiomers to be more easily and accurately described. The same year saw first successful enantiomeric separation by gas chromatography an important development as the technology was in common use at the time.

Metal catalysed enantioselective synthesis was pioneered by William S. Knowles, Ryō-ji Noyori and K. Barry Sharpless; for which they would receive the 2001 Nobel Prize in Chemistry. Knowles and Noyori began with the development of asymmetric hydrogenation, which they developed independently in 1968. Knowles replaced the achiral triphenylphosphine ligands in Wilkinson's catalyst with chiral phosphine ligands. This experimental catalyst was employed in an asymmetric hydrogenation with a modest 15% enantiomeric excess. Knowles was also the first to apply enantioselective metal catalysis to industrial-scale synthesis; while working for the Monsanto Company he developed an enantioselective hydrogenation step for the production of L-DOPA, utilising the DIPAMP ligand.

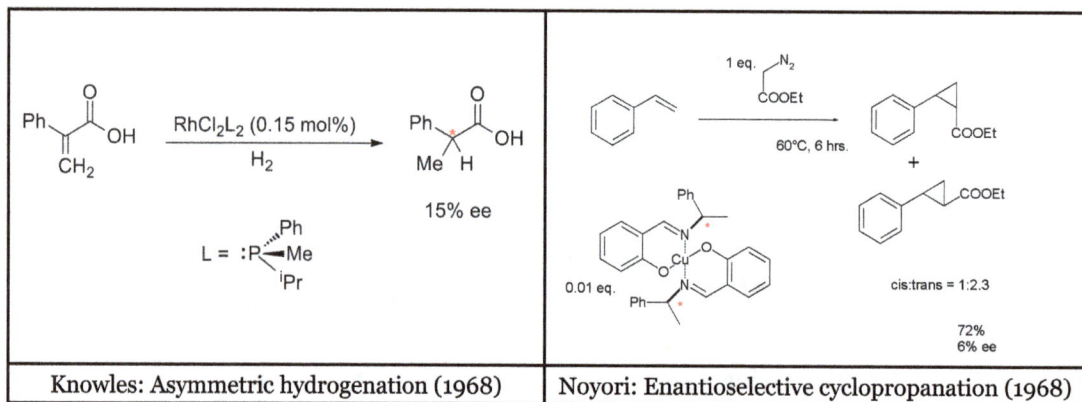

| Knowles: Asymmetric hydrogenation (1968) | Noyori: Enantioselective cyclopropanation (1968) |

Noyori devised a copper complex using a chiral Schiff base ligand, which he used for the metal-carbenoid cyclopropanation of styrene. In common with Knowles' findings, Noyori's results for the enantiomeric excess for this first-generation ligand were disappointingly low: 6%. However continued research eventually led to the development of the Noyori asymmetric hydrogenation reaction.

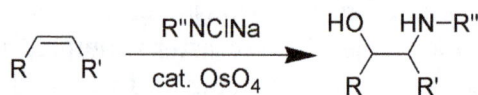

The Sharpless oxyamination

Sharpless complemented these reduction reactions by developing a range of asymmetric oxidations (Sharpless epoxidation, Sharpless asymmetric dihydroxylation, Sharpless oxyamination) during the 1970s to 1980's. With the asymmetric oxyamination reaction, using osmium tetroxide, being the earliest.

During the same period, methods were developed to allow the analysis of chiral compounds by NMR; either using chiral derivatizing agents, such as Mosher's acid, or europium based shift reagents, of which $Eu(DPM)_3$ was the earliest.

Chiral auxiliaries were introduced by E.J. Corey in 1978 and featured prominently in the work of Dieter Enders. Around the same time enantioselective organocatalysis was developed, with pioneering work including the Hajos–Parrish–Eder–Sauer–Wiechert reaction. Enzyme-catalyzed enantioselective reactions became more and more common during the 1980s, particularly in industry, with their applications including asymmetric ester hydrolysis with pig-liver esterase. The emerging technology of genetic engineering has allowed the tailoring of enzymes to specific processes, permitting an increased range of selective transformations. For example, in the asymmetric hydrogenation of statin precursors.

References

- Albericio, F. (2000). Solid-Phase Synthesis: A Practical Guide (1 ed.). Boca Raton: CRC Press. p. 848. ISBN 0-8247-0359-6.

- Atherton, E.; Sheppard, R.C. (1989). Solid Phase peptide synthesis: a practical approach. Oxford, England: IRL Press. ISBN 0-19-963067-4.

- Stewart, J.M.; Young, J.D. (1984). Solid phase peptide synthesis (2nd ed.). Rockford: Pierce Chemical Company. p. 91. ISBN 0-935940-03-0.

- Manchester KL (1964). "Sites of Hormonal Regulation of Protein Metabolism". In Munro HN, Allison JB. Mammalian protein metabolism. 4. New York: Academic Press. p. 229. ISBN 978-0-12-510604-7.

- White D (2007). The physiology and biochemistry of prokaryotes (3rd ed.). New York: Oxford Univ. Press. ISBN 0195301684.

- Figge RM (2007). "Methione biosynthesis". In Wendisch VF. Amino acid biosynthesis: pathways, regulation, and metabolic engineering. Berlin: Springer. pp. 206–208. ISBN 3540485953.

- Lehninger AL, Cox MM, Nelson DL (2008). Lehninger principles of biochemistry (5th ed.). New York: W.H. Freeman. p. 528. ISBN 978-0-7167-7108-1.

- Berg JM, Tymoczko JL, Stryer L (2002). Biochemistry (5th ed.). New York, NY: W. H. Freeman. ISBN 0-7167-3051-0.

- Leo A. Paquette :Chiral Reagents for Asymmetric Synthesis, S.220-223, 2003, Wiley and Sons, ISBN 0470856254

- Jan Bülle, Aloys Hüttermann : Das Basiswissen der organischen Chemie: Die wichtigsten organischen Reaktionen im Labor und in der Natur, S.310/311, 2000, Wiley-VCH, ISBN 3527308474

- Clayden, Jonathan; Greeves, Nick; Warren, Stuart; Wothers, Peter (2001). Organic Chemistry (1st ed.). Oxford University Press. ISBN 978-0-19-850346-0. Page 1226

- N. Jacobsen, Eric; Pfaltz, Andreas; Yamamoto, Hisashi (1999). Comprehensive asymmetric catalysis 1-3. Berlin: Springer. ISBN 9783540643371.

- Roos, Gregory (2002). Compendium of chiral auxiliary applications. San Diego, Calif. [u.a.]: Acad. Press. ISBN 9780125953443.

- Evans, D. A.; Helmchen, G.; Rüping, M. (2007). "Chiral Auxiliaries in Asymmetric Synthesis". In Christmann, M. Asymmetric Synthesis – The Essentials. Wiley-VCH Verlag GmbH & Co. pp. 3–9. ISBN 978-3-527-31399-0.

- Faber, Kurt (2011). Biotransformations in organic chemistry a textbook (6th rev. and corr. ed.). Berlin: Springer-Verlag. ISBN 9783642173936.

- Gröger, Albrecht Berkessel; Harald (2005). Asymmetric organocatalysis – from biomimetic concepts to applications in asymmetric synthesis (1. ed., 2. reprint. ed.). Weinheim: Wiley-VCH. ISBN 3-527-30517-3.

- Koskinen, Ari M.P. (2013). Asymmetric synthesis of natural products (Second ed.). Hoboken, N.J.: Wiley. pp. 17, 28–29. ISBN 1118347331

Expanded Genetic Code: An Integrated Study

Expanded genetic codes are unnaturally altered genetic codes. Expanding genetic code is an interdisciplinary subject of synthetic biology. The important topics related to expanded genetic coding are peptide synthesis, genetic code, transfer RNA and stop codon. This chapter elucidates the crucial theories and principles relater to expanded genetic code.

Expanded Genetic Code

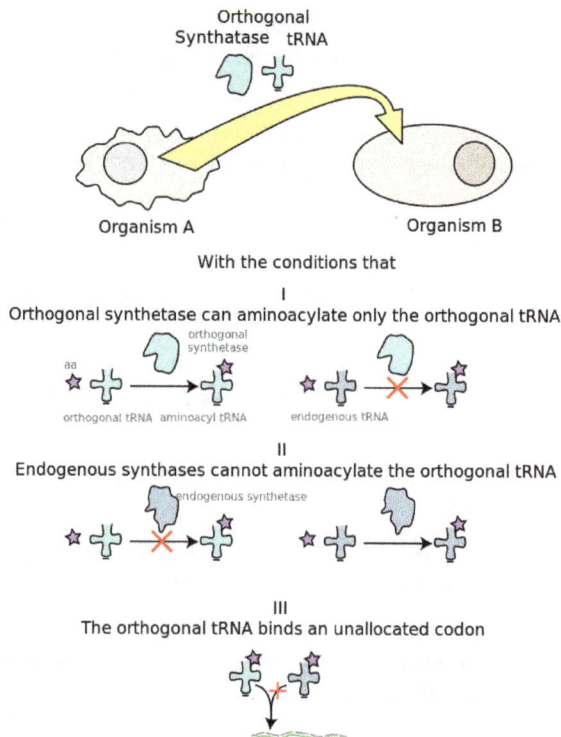

Orthogonal
Synthatase   tRNA

Organism A   Organism B

With the conditions that

I
Orthogonal synthetase can aminoacylate only the orthogonal tRNA

orthogonal
synthetase

aa

orthogonal tRNA aminoacyl tRNA   endogenous tRNA

II
Endogenous synthases cannot aminoacylate the orthogonal tRNA

endogenous synthetase

III
The orthogonal tRNA binds an unallocated codon

There must not be crosstalk between the new tRNA/synthase pair and the existing tRNA/synthase molecules, only with the ribosomes

An expanded genetic code is an artificially modified genetic code in which one or more specific codons have been re-allocated to encode an amino acid that is not among the 20 encoded proteinogenic amino acids.

The key prerequisites to expand the genetic code are:

- the non-standard amino acid to encode,

- an unused codon to adopt,

- a tRNA that recognises this codon, and

- a tRNA synthetase that recognises only that tRNA and only the non-standard amino acid.

Expanding the genetic code is an area of research of synthetic biology, an applied biological discipline whose goal is to engineer living systems for useful purposes. The genetic code expansion enriches the repertoire of useful tools available to science.

Introduction

Proteins are produced thanks to the translational system molecules, which decode the RNA messages into a string of amino acids. The translation of genetic information contained in messenger RNA (mRNA) into a protein is catalysed by ribosomes. Transfer RNAs (tRNA) are used as keys to decode the mRNA into its encoded polypeptide. The tRNA recognizes a specific three nucleotide codon in the mRNA with a complementary sequence called the anticodon on one of its loops. Each three nucleotide codon is translated into one of twenty naturally occurring amino acids. There is at least one tRNA for any codon, and sometimes multiple codons code for the same amino acid. Many tRNAs are compatible with several codons. An enzyme called an aminoacyl tRNA synthetase covalently attaches the amino acid to the appropriate tRNA. Most cells have a different synthetase for each amino acid (20 or more synthetases). On the other hand, some bacteria have fewer than 20 aminoacyl tRNA synthetases, and introduce the "missing" amino acid(s) by modification of a structurally related amino acid by an aminotransferase enzyme. A feature exploited in the expansion of the genetic code is the fact the aminoacyl tRNA synthetase often does not recognize the anticodon, but another part of the tRNA, meaning that if the anticodon were to be mutated the encoding of that amino acid would change to a new codon. In the ribosome, the information in mRNA is translated into a specific amino acid when the mRNA codon matches with the complementary anticodon of a tRNA, and the attached amino acid is added onto a growing polypeptide chain. When it is released from the ribosome, the polypeptide chain folds into a functioning protein.

In order to incorporate a novel amino acid into the genetic code several changes are required. First, for successful translation of a novel amino acid, the codon to which the novel amino acid is assigned cannot already code for one of the 20 natural amino acids. Usually a nonsense codon (stop codon) or a four-base codon are used. Second, a novel pair of tRNA and aminoacyl tRNA synthetase are required, these are called the orthogonal set. The orthogonal set must not crosstalk with the endogenous tRNA and synthetase sets, while still being functionally compatible with the ribosome and other components of the translation apparatus. The active site of the synthetase is modified to accept only the novel amino acid. Most often, a library of mutant synthetases is screened for one which charges the tRNA with the desired amino acid. The synthetase is also modified to recognize only the orthogonal tRNA. The tRNA synthetase pair is often engineered in other bacteria or eukaryotic cells.

In this area of research, the 20 encoded proteinogenic amino acids are referred to as standard amino acids, or alternatively as natural or canonical amino acids, while the added amino acids are called non-standard amino acids (NSAAs), or unnatural amino acids (uAAs; term not used in papers dealing with natural non-proteinogenic amino acids, such as phosphoserine), or non-canonical amino acids.

Non-standard Amino Acids

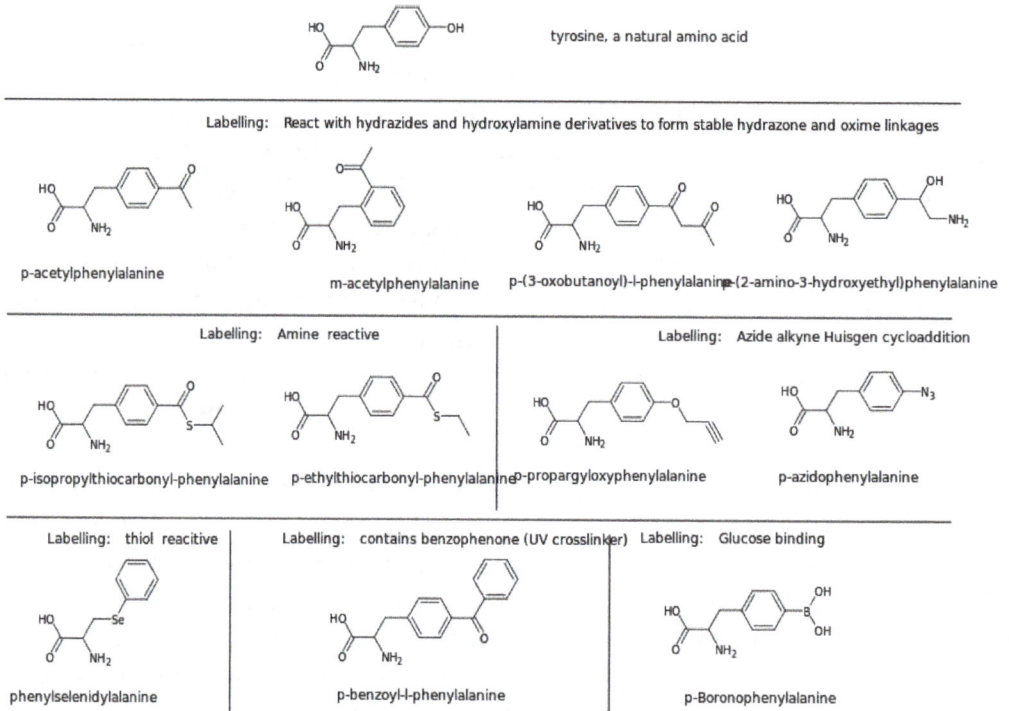

Some amino acids that have been added in order to label protein

The first element of the system is the amino acid that is added to the genetic code of an certain strain of organism.

Over 71 different NSAAs have been added to different strains of *E. coli*, yeast or mammalian cells. Due to technical details (easier chemical synthesis of NSAAs, less crosstalk and easier evolution of the aminoacyl-tRNA synthase), the NSAAs are generally larger than standard amino acids and most often have a phenylalanine core but with a large variety of different substitutents. These allow a large repertoire of new functions, such as labelling, as a fluorescent reporter (*e.g.* dansylalanine) or to produce translationally protein in *E. coli* with Eukaryotic post-translational modifications (*e.g.* phosphoserine, phosphothreonine, and phosphotyrosine).

Unnatural amino acids incorporated into proteins include heavy atom containing amino acids to facilitate x-ray crystallographic studies; amino acids with novel steric/packing and electronic properties; photocrosslinking amino acids which can be used to probe protein-protein interactions in vitro or in vivo; keto, acetylene, azide, and boronate containing amino acids which can be used to selectively introduce a large number of biophysical probes, tags, and novel chemical functional groups into proteins *in vitro* or *in vivo*; redox active amino acids to probe and modulate electron transfer; photocaged and photoisomerizable amino acids to photoregulate biological processes; metal binding amino acids for catalysis and metal ion sensing; amino acids that contain fluorescent or infra-red active side chains to probe protein structure and dynamics; α-hydroxy acids and D-amino acids as probes of backbone conformation and hydrogen bonding interactions; and sulfated amino acids and mimetics of phosphorylated amino acids as probes of posttranslational modifications.

Availability of the non-standard amino acid requires that the organism either import it from the medium or biosynthesised it. In the first case, the unnatural amino acid is first synthesised chemically in optically pure L-form. It is then added to the growth medium of the cell. Generally a library of compounds is tested to see which can be imported and incorporated, but often the various transport systems can handle unnatural amino acids with apolar side-chains. In the second case, a biosynthetic paths need to be engineered. One example is an *E. coli* strain that biosynthesizes a novel, previously unnatural amino acid (p-aminophenylalanine) from basic carbon sources and includes this amino acid in its genetic code. Another example is the production of phosphoserine, which is a natural metabolite and as a consequence the pathway flux had to be altered to increase its production.

Codon Assignment

Another element of the system is a codon to allocate to the new amino acid.

A major problem for the genetic code expansion is that there are no free codons. The genetic code has a nonrandom layout that shows tell-tale signs of various phases of primordial evolution, however, it has since frozen into place and is near-universally conserved. Nevertheless, some codons are rarer than others. In fact, in *E. coli* (and all organisms) the codon usage is not equal, but presents several rare codons (see table), the rarest being the amber stop codon (UAG).

Codon usage in E. coli		
Codon	**Amino acid**	**% abundance**
UUU	Phe (F)	1.9
UUC	Phe (F)	1.8
UUA	Leu (L)	1.0
UUG	Leu (L)	1.1
CUU	Leu (L)	1.0
CUC	Leu (L)	0.9
CUA	Leu (L)	0.3
CUG	Leu (L)	5.2
AUU	Ile (I)	2.7
AUC	Ile (I)	2.7
AUA	Ile (I)	0.4
AUG	Met (M)	2.6
GUU	Val (V)	2.0
GUC	Val (V)	1.4
GUA	Val (V)	1.2
GUG	Val (V)	2.4
UCU	Ser (S)	1.1
UCC	Ser (S)	1.0
UCA	Ser (S)	0.7
UCG	Ser (S)	0.8
CCU	Pro (P)	0.7
CCC	Pro (P)	0.4
CCA	Pro (P)	0.8
CCG	Pro (P)	2.4
ACU	Thr (T)	1.2
ACC	Thr (T)	2.4
ACA	Thr (T)	0.1
ACG	Thr (T)	1.3
GCU	Ala (A)	1.8
GCC	Ala (A)	2.3
GCA	Ala (A)	0.1
GCG	Ala (A)	3.2
UAU	Tyr (Y)	1.6
UAC	Tyr (Y)	1.4
UAA	Stop	0.2
UAG	Stop	0.03
CAU	His (H)	1.2
CAC	His (H)	1.1

Codon usage in E. coli		
Codon	**Amino acid**	**% abundance**
CAA	Gln (Q)	1.3
CAG	Gln (Q)	2.9
AAU	Asn (N)	1.6
AAC	Asn (N)	2.6
AAG	Lys (K)	3.8
AAA	Lys (K)	1.2
GAU	Asp (D)	3.3
GAC	Asp (D)	2.3
GAA	Glu (E)	4.4
GAG	Glu (E)	1.9
UGU	Cys (C)	0.4
UGC	Cys (C)	0.6
UGA	Stop	0.01
UGG	Trp (W)	1.4
CGU	Arg (R)	2.4
CGC	Arg (R)	2.2
CGA	Arg (R)	0.3
CGG	Arg (R)	0.5
AGU	Ser (S)	0.7
AGC	Ser (S)	1.5
AGA	Ser (S)	0.2
AGG	Ser (S)	0.2
GGU	Gly (G)	2.8
GGC	Gly (G)	3.0
GGC	Gly (G)	0.7
GGA	Gly (G)	0.9

Amber Codon Suppression

The possibility of reassigning codons was realized by Normanly *et al.* in 1990, when a viable mutant strain of *E. coli* read through the UAG ("amber") stop codon. This was possible thanks to the rarity of this codon and the fact that release factor 1 alone makes the amber codon terminate translation. Later, in the Schultz lab, the tRNATyr/ tyrosyl-tRNA synthetase (TyrRS) from *Methanococcus jannaschii*, an archaebacterium, was used to introduce a tyrosine instead of STOP, the default value of the amber codon. This was possible because of the differences between the endogenous bacterial synthases and the orthologous archaeal synthase, which do not recognize each other. Subsequently, the group evolved the orthologonal tRNA/synthase pair to utilise the non-standard amino acid *O*-methyltyrosine. This was followed by the larger naphthyl-alanine and the photocrosslinking benzoylphenylalanine, which proved the potential utility of the system.

The amber codon is the least used codon in *Escherichia coli*, but its highjacking results in a substantial loss of fitness. One study in fact found that there were at least 83 peptides majorly affected by the readthrough Additionally, the labelling was incomplete. As a consequence, several strains have been made to reduce the fitness cost, including the removal of all amber codons from the genome. In most *E. coli* K-12 strains (viz. *Escherichia coli* (molecular biology) for strain pedigrees) there are 314 UAG stop codons. Consequently, a gargantuan amount of work has gone into the replacement of these. One approach pioneered by the group of Prof. George Church from Harvard, was dubbed MAGE in CAGE: this relied on a multiplex transformation and subsequent strain recombination to remove all UAG codons—the latter part presented a halting point in a first paper, but was overcome. This resulted in the *E. coli* strain C321.ΔA, which lacks all UAG codons and RF1. This allowed an experiment to be done with this strain to make it "addicted" to the amino acid biphenylalanine by evolving several key enzymes to require it structurally, therefore putting its expanded genetic code under positive selection.

Rare Sense Codon Reassignment

In addition to the amber codon, rare sense codons have also been considered for use. The AGG codon codes for arginine, but a strain has been successfully modified to make it code for 6-*N*-allyloxycarbonyl-lysine. Another candidate is the AUA codon, which is unusual in that its respective tRNA has to differentiate against AUG that codes for methionine (primordially, isoleucine, hence its location). In order to do this, the AUA tRNA has a special base, lysidine. The deletion of the synthase (*tilS*) was possible thanks to the replacement of the native tRNA with that of *Mycoplasma mobile* (no lysidine). The reduced fitness is a first step towards pressuring the strain to loose all instances of AUA, allowing it to be used for genetic code expansion.

Four Base Codons

Other approaches include the addition of extra base pairing or the use of orthologous ribosomes that accept in addition to the regular triplet genetic code, tRNAs with quadruple code. This allowed the simultaneous usage of two unnatural amino acids, *p*-azidophenylalanine (AzPhe) and N6-[(2-propynyloxy)carbonyl]lysine (CAK), which cross-link with each other by Huisgen cycloaddition.

tRNA/Synthase pair

Another key element is the tRNA/synthase pair.

The orthologous set of synthetase and tRNA can be mutated and screened through directed evolution to charge the tRNA with a different, even novel, amino acid. Mutations to the plasmid containing the pair can be introduced by error-prone PCR or through degenerate primers for the synthetase's active site. Selection involves multiple rounds of a two-step process, where the plasmid is transferred into cells expressing chloramphenicol acetyl transferase with a premature amber codon. In the presence of toxic chloramphenicol and the non-natural amino acid, the surviving cells will have overridden the amber codon using the orthogonal tRNA aminoacylated with either the standard amino acids or the non-natural one. To remove the former, the plasmid is inserted into cells with a barnase gene (toxic) with a premature amber codon but without the non-natural amino acid, removing all the orthogonal synthases that do not specifically recognize the non-natural amino acid. In addition to the recoding of the tRNA to a different codon, they can be mutated to recognize a four-base codon, allowing additional free coding options. The non-natural amino acid, as a result, introduces diverse physicochemical and biological properties in order to be used as a tool to explore protein structure and function or to create novel or enhanced protein for practical purposes.

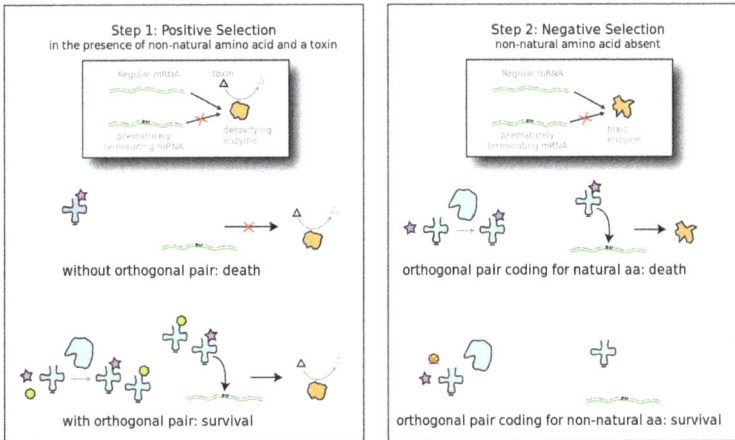

Several methods for selecting the synthetase that accepts only the non-natural amino acid have been developed. One of which is by using a combination of positive and negative selection

Orthogonal Sets in *E. coli*

The orthogonal pairs of synthetase and tRNA that work for one organism may not work for another, as the synthetase may mis-aminoacylate endogenous tRNAs or the tRNA be mis-aminoacylated itself by an endogenous synthetase. As a result, the sets created to date differ between organisms.

- tRNA^{Tyr}-TyrRS pair from the archaeon *Methanococcus jannaschii*

- tRNA^{Lys}–LysRS pair from the archaeon *Pyrococcus horikoshii*

- tRNA^{Glu}–GluRS pair from *Methanosarcina mazei*

- leucyl-tRNA synthetase from *Methanobacterium thermoautotrophicum* and a mutant leucyl tRNA derived from *Halobacterium* sp

- tRNA^Amber^-PylRS pair from the archaeon *Methanosarcina barkeri* and *Methanosarcina mazei*

- tRNA^Amber^-3-Iodotyrosyl-tRNA synthetase (variant of aminoacyl-tRNA synthetase from *Methanocaldococcus jannaschii*)

Orthogonal Sets in yeast

- tRNA^Tyr^-TyrRS pair from *Escherichia coli*

- tRNA^Leu^–LeuRS pair from *Escherichia coli*

- tRNA^iMet^ from human and GlnRS from *Escherichia coli*

- tRNA^Amber^-PylRS pair from the archaeon *Methanosarcina barkeri* and *Methanosarcina mazei*

- tRNA^Amber^-4,5-dimethoxy-2-nitrobenzyl-cysteinyl-tRNA synthetase

Orthogonal Sets in Mammalian Cells

- tRNA^Tyr^-TyrRS pair from *Bacillus stearothermophilus*

- modified tRNA^Trp^-TrpRS pair from *Bacillus subtilis* trp

- tRNA^Leu^–LeuRS pair from *Escherichia coli*

- tRNA^Amber^-PylRS pair from the archaeon *Methanosarcina barkeri* and *Methanosarcina mazei*

Applications

With an expanded genetic code, the unnatural amino acid can be genetically directed to any chosen site in the protein of interest. The high efficiency and fidelity of this process allows a better control of the placement of the modification compared to modifying the protein post-translationally, which, in general, will target all amino acids of the same type, such as the thiol group of cysteine and the amino group of lysine. Also, an expanded genetic code allows modifications to be carried out *in vivo*. The ability to site-specifically direct lab-synthesized chemical moieties into proteins allows many types of studies that would otherwise be extremely difficult, such as:

- Probing protein structure and function: By using amino acids with slightly different size such as *O*-methyltyrosine or dansylalanine instead of tyrosine, and by inserting genetically coded reporter moieties (color-changing and/or spin-active) into selected protein sites, chemical information about the protein's structure and function can be measured.

- Probing the role of post-translational modifications in protein structure and function: By using amino acids that mimic post-translational modifications such as phosphoserine, biologically active protein can be obtained, and the site-specific nature of the amino acid incorporation can lead to information on how the position, density, and distribution of protein phosphorylation effect protein function.

- Identifying and regulating protein activity: By using photocaged aminoacids, protein function can be "switched" on or off by illuminating the organism.

- Changing the mode of action of a protein: One can start with the gene for a protein that binds a certain sequence of DNA and, by inserting a chemically active amino acid into the binding site, convert it to a protein that cuts the DNA rather than binding it.

- Improving immunogenicity and overcoming self-tolerance: By replacing strategically chosen tyrosines with p-nitro phenylalanine, a tolerated self-protein can be made immunogenic.

- Selective destruction of selected cellular components: using an expanded genetic code, unnatural, destructive chemical moieties (sometimes called "chemical warheads") can be incorporated into proteins that target specific cellular components.

- Producing better protein: the evolution of T7 bacteriophages on a non-evolving *E. coli* strain that encoded 3-iodotyrosine on the amber codon, resulted in a population fitter than wild-type thanks to the presence of iodotyrosine in its proteome

Future

The expansion of the genetic code is still in its infancy. Current methodology uses only one non-standard amino acid at the time, whereas ideally multiple could be used.

Recoded Synthetic Genome

One way to achieve the encoding of multiple unnatural amino acids is by rewriting genome synthetically. In 2010, at the cost of $40 million an organism, *Mycoplasma laboratorium*, was constructed that was controlled by a synthetic genome. Due to the larger genome size this is not possible with *E. coli*, however several methods are being developed to overcome this, such as the fragmentation of the genome into separate linear chromosomes. In addition to the elimination of the usage of rare codons, the specificity of the system needs to be increased as many tRNA recognise several codons

Expanded Genetic Alphabet

Another approach is to expand the number of nucleobases to increase the coding capacity.

An unnatural base pair (UBP) is a designed subunit (or nucleobase) of DNA which is created in a laboratory and does not occur in nature. A demonstration of UBPs were achieved *in vitro* by Ichiro Hirao's group at RIKEN institute in Japan. In 2002, they developed an unnatural base pair between 2-amino-8-(2-thienyl)purine (s) and pyridine-2-one (y) that functions *in vitro* in transcription and translation for the site-specific incorporation of non-standard amino acids into proteins. In 2006, they created 7-(2-thienyl)imidazo[4,5-b] pyridine (Ds) and pyrrole-2-carbaldehyde (Pa) as a third base pair for replication and transcription. Afterward, Ds and 4-[3-(6-aminohexanamido)-1-propynyl]-2-nitropyrrole (Px) was discovered as a high fidelity pair in PCR amplification. In 2013, they applied the Ds-Px pair to DNA aptamer generation by *in vitro* selection (SELEX) and demonstrated the genetic alphabet expansion significantly augment DNA aptamer affinities to target proteins.

In 2012, a group of American scientists led by Floyd Romesberg, a chemical biologist at the Scripps Research Institute in San Diego, California, published that his team designed an unnatural base pair (UBP). The two new artificial nucleotides or *Unnatural Base Pair* (UBP) were named d5SICS and dNaM. More technically, these artificial nucleotides bearing hydrophobic nucleobases, feature two fused aromatic rings that form a (d5SICS–dNaM) complex or base pair in DNA. In 2014 the same team from the Scripps Research Institute reported that they synthesized a stretch of circular DNA known as a plasmid containing natural T-A and C-G base pairs along with the best-performing UBP Romesberg's laboratory had designed, and inserted it into cells of the common bacterium *E. coli* that successfully replicated the unnatural base pairs through multiple generations. This is the first known example of a living organism passing along an expanded genetic code to subsequent generations. This was in part achieved by the addition of a supportive algal gene that expresses a nucleotide triphosphate transporter which efficiently imports the triphosphates of both d5SICSTP and dNaMTP into *E. coli* bacteria. Then, the natural bacterial replication pathways use them to accurately replicate the plasmid containing d5SICS–dNaM.

The successful incorporation of a third base pair into a living micro-organism is a significant breakthrough toward the goal of greatly expanding the number of amino acids which can be encoded by DNA, thereby expanding the potential for living organisms to produce novel proteins. The artificial strings of DNA do not encode for anything yet, but scientists speculate they could be designed to manufacture new proteins which could have industrial or pharmaceutical uses.

In May 2014, researchers announced that they had successfully introduced two new artificial nucleotides into bacterial DNA, and by including individual artificial nucleotides in the culture media, were able to passage the bacteria 24 times; they did not create mRNA or proteins able to use the artificial nucleotides.

Related Methods

Alloprotein

There have been many studies that have produced protein with non-standard amino acids, but they do not alter the genetic code. These protein, called alloprotein, are made by incubating cells with an unnatural amino acid in the absence of a similar coded amino acid in order for the former to be incorporated into protein in place of the latter, for example L-2-aminohexanoic acid (Ahx) for methionine (Met).

These studies rely on the natural promiscuous activity of the amino-acyl synthase to add to its target tRNA an unnatural amino acid similar to the natural substrate, for example methionyl-tRNA synthase's mistaking isoleucine for methionine. In protein crystallography, for example, the addition of selenomethionine to the media of a culture of a methionine-auxotrophic strain results in proteins containing selenomethionine as opposed to methionine (*viz.* Multi-wavelength anomalous dispersion for reason). Another example is that photoleucine and photomethionine are added instead of leucine and methionine to cross-label protein. Similarly, some tellurium-tolerant fungi can incorporate tellurocysteine and telluromethionine into their protein instead of cysteine and methionine. The objective of expanding the genetic code is more radical as it does not replace an amino acid, but it adds one or more to the code.

In Vitro Synthesis

The genetic code expansion described above is *in vivo*. An alternative is the change of coding *in vitro* translation experiments. This requires the depletion of all tRNAs and the selective reintroduction of certain aminoacylated-tRNAs, some chemically aminoacylated.

Chemical Synthesis

There are several techniques to produce peptides chemically, generally it is by solid-phase protection chemistry. This means that any (protected) amino acid can be added into the nascent sequence.

Genetic Code

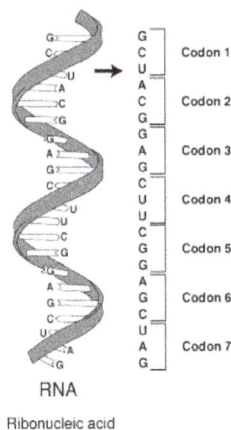

RNA

Ribonucleic acid

A series of codons in part of a messenger RNA (mRNA) molecule. Each codon consists of three nucleotides, usually corresponding to a single amino acid. The nucleotides are abbreviated with the letters A, U, G and C. This is mRNA, which uses U (uracil). DNA uses T (thymine) instead. This mRNA molecule will instruct a ribosome to synthesize a protein according to this code.

The genetic code is the set of rules by which information encoded within genetic material (DNA or mRNA sequences) is translated into proteins by living cells. Translation is accomplished by the ribosome, which links amino acids in an order specified by mRNA, using transfer RNA (tRNA) molecules to carry amino acids and to read the mRNA three nucleotides at a time. The genetic code is highly similar among all organisms and can be expressed in a simple table with 64 entries.

The code defines how sequences of nucleotide triplets, called *codons*, specify which amino acid will be added next during protein synthesis. With some exceptions, a three-nucleotide codon in a nucleic acid sequence specifies a single amino acid. Because the vast majority of genes are encoded with exactly the same code, this particular code is often referred to as the canonical or standard genetic code, or simply *the* genetic code, though in fact some variant codes have evolved. For example, protein synthesis in human mitochondria relies on a genetic code that differs from the standard genetic code.

While the "genetic code" determines a protein's amino acid sequence, other genomic regions determine when and where these proteins are produced according to a multitude of more complex "gene regulatory codes".

Discovery

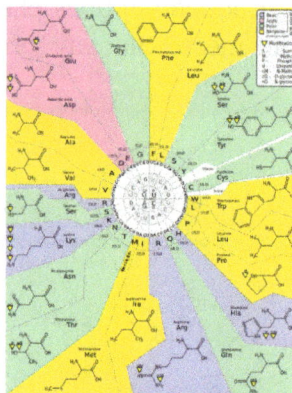

The genetic code

Serious efforts to understand how proteins are encoded began after the structure of DNA was discovered in 1953. George Gamow postulated that sets of three bases must be employed to encode the 20 standard amino acids used by living cells to build proteins. With four different nucleotides, a code of 2 nucleotides would allow for only a maximum of $4^2 = 16$ amino acids. A code of 3 nucleotides could code for a maximum of $4^3 = 64$ amino acids.

The Crick, Brenner et al. experiment first demonstrated that codons consist of three DNA bases; Marshall Nirenberg and Heinrich J. Matthaei were the first to elucidate the nature of a codon in 1961 at the National Institutes of Health. They used a cell-free system to translate a poly-uracil RNA sequence (i.e., UUUUU...) and discovered that the polypeptide that they had synthesized consisted of only the amino acid phenylalanine. They thereby deduced that the codon UUU specified the amino acid phenylalanine. This was followed by experiments in Severo Ochoa's laboratory that demonstrated that the poly-adenine RNA sequence (AAAAA...) coded for the polypeptide poly-lysine and that the poly-cytosine RNA sequence (CCCCC...) coded for the polypeptide poly-proline. Therefore, the codon AAA specified the amino acid lysine, and the codon CCC specified the amino acid proline. Using different copolymers most of the remaining codons were then determined. Subsequent work by Har Gobind Khorana identified the rest of the genetic code. Shortly thereafter, Robert W. Holley determined the structure of transfer RNA (tRNA), the adapter molecule that facilitates the process of translating RNA into protein. This work was based upon earlier studies by Severo Ochoa, who received the Nobel Prize in Physiology or Medicine in 1959 for his work on the enzymology of RNA synthesis.

Extending this work, Nirenberg and Philip Leder revealed the triplet nature of the genetic code and deciphered the codons of the standard genetic code. In these experiments, various combinations of mRNA were passed through a filter that contained ribosomes, the components of cells that translate RNA into protein. Unique triplets promoted the binding of specific tRNAs to the ribosome. Leder and Nirenberg were able to determine the sequences of 54 out of 64 codons in their experiments. In 1968, Khorana, Holley and Nirenberg received the Nobel Prize in Physiology or Medicine for their work.

Features

Reading Frame

A codon is defined by the initial nucleotide from which translation starts and sets the frame for a run of uninterrupted triplets, which is known as an "open reading frame" (ORF). For example, the string GGGAAACCC, if read from the first position, contains the codons GGG, AAA, and CCC; and, if read from the second position, it contains the codons GGA and AAC; if read starting from the third position, GAA and ACC. Every sequence can, thus, be read in its 5' → 3' direction in three reading frames, each of which will produce a different amino acid sequence (in the given example, Gly-Lys-Pro, Gly-Asn, or Glu-Thr, respectively). With double-stranded DNA, there are six possible reading frames, three in the forward orientation on one strand and three reverse on the opposite strand. The actual frame from which a protein sequence is translated is defined by a start codon, usually the first AUG codon in the mRNA sequence.

In eukaryotes, ORFs in exons are often interrupted by introns.

Start/Stop Codons

Translation starts with a chain initiation codon or start codon. Unlike stop codons, the codon alone is not sufficient to begin the process. Nearby sequences such as the Shine-Dalgarno sequence in *E. coli* and initiation factors are also required to start translation. The most common start codon is AUG, which is read as methionine or, in bacteria, as formylmethionine. Alternative start codons depending on the organism include "GUG" or "UUG"; these codons normally represent valine and leucine, respectively, but as start codons they are translated as methionine or formylmethionine.

The three stop codons have been given names: UAG is *amber*, UGA is *opal* (sometimes also called *umber*), and UAA is *ochre*. "Amber" was named by discoverers Richard Epstein and Charles Steinberg after their friend Harris Bernstein, whose last name means "amber" in German. The other two stop codons were named "ochre" and "opal" in order to keep the "color names" theme. Stop codons are also called "termination" or "nonsense" codons. They signal release of the nascent polypeptide from the ribosome because there is no cognate tRNA that has anticodons complementary to these stop signals, and so a release factor binds to the ribosome instead.

Effect of Mutations

During the process of DNA replication, errors occasionally occur in the polymerization of the second strand. These errors, called mutations, can affect the phenotype of an organism, especially if they occur within the protein coding sequence of a gene. Error rates are usually very low—1 error in every 10–100 million bases—due to the "proofreading" ability of DNA polymerases.

Missense mutations and nonsense mutations are examples of point mutations, which can cause genetic diseases such as sickle-cell disease and thalassemia respectively. Clinically important missense mutations generally change the properties of the coded amino acid residue between being basic, acidic, polar or non-polar, whereas nonsense mutations result in a stop codon.

Mutations that disrupt the reading frame sequence by indels (insertions or deletions) of a non-multiple of 3 nucleotide bases are known as frameshift mutations. These mutations usually result in a completely different translation from the original, and are also very likely to cause a stop codon to be read, which truncates the creation of the protein. These mutations may impair the function of the resulting protein, and are thus rare in *in vivo* protein-coding sequences. One reason inheritance of frameshift mutations is rare is that, if the protein being translated is essential for growth under the selective pressures the organism faces, absence of a functional protein may cause death before the organism is viable. Frameshift mutations may result in severe genetic diseases such as Tay-Sachs disease.

Although most mutations that change protein sequences are harmful or neutral, some mutations have a beneficial effect on an organism. These mutations may enable the mutant organism to withstand particular environmental stresses better than wild type organisms, or reproduce more quickly. In these cases a mutation will tend to become more common in a population through natural selection. Viruses that use RNA as their genetic material have rapid mutation rates, which can be an advantage, since these viruses will evolve constantly and rapidly, and thus evade the defensive responses of e.g. the human immune system. In large populations of asexually reproducing organisms, for example, *E. coli*, multiple beneficial mutations may co-occur. This phenomenon is called clonal interference and causes competition among the mutations.

Degeneracy

Degeneracy is the redundancy of the genetic code. This term was given by Bernfield and Nirenberg. The genetic code has redundancy but no ambiguity (see the codon tables below for the full correlation). For example, although codons GAA and GAG both specify glutamic acid (redundancy), neither of them specifies any other amino acid (no ambiguity). The codons encoding one amino acid may differ in any of their three positions. For example, the amino acid leucine is specified by YUR or CUN (UUA, UUG, CUU, CUC, CUA, or CUG) codons (difference in the first or third position indicated using IUPAC notation), while the amino acid serine is specified by UCN or AGY (UCA, UCG, UCC, UCU, AGU, or AGC) codons (difference in the first, second, or third position). A practical consequence of redundancy is that errors in the third position of the triplet codon cause only a silent mutation or an error that would not affect the protein because the hydrophilicity or hydrophobicity is maintained by equivalent substitution of amino acids; for example, a codon of NUN (where N = any nucleotide) tends to code for hydrophobic amino acids. NCN yields amino acid residues that are small in size and moderate in hydropathy; NAN encodes average size hydrophilic residues. The genetic code is so well-structured for hydropathy that a mathematical analysis (Singular Value Decomposition) of 12 variables (4 nucleotides x 3 positions) yields a remarkable correlation (C = 0.95) for predicting the hydropathy of the encoded amino acid directly from the triplet nucleotide sequence, *without translation*. Note in the table, below, eight amino acids are not affected at all by mutations at the third position of the codon, whereas in the figure above, a mutation at the second position is likely to cause a radical change in the physicochemical properties of the encoded amino acid.

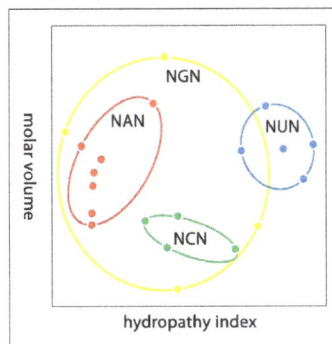

Grouping of codons by amino acid residue molar volume and hydropathy. A more detailed version is available.

Codon Usage Bias

The frequency of codons, also known as codon usage bias, can vary from species to species with functional implications for the control of translation. The following codon usage table is for the human genome.

Human genome codon frequency

Codon	AA	Fraction	Freq ‰	Number	Codon	AA	Fraction	Freq ‰	Number	Codon	AA	Fraction	Freq ‰	Number	Codon	AA	Fraction	Freq ‰	Number
UUU	F	0.46	17.6	714298	UCU	S	0.19	15.2	618711	UAU	Y	0.44	12.2	495699	UGU	C	0.46	10.6	430311
UUC	F	0.54	20.3	824692	UCC	S	0.22	17.7	718892	UAC	Y	0.56	15.3	622407	UGC	C	0.54	12.6	513028
UUA	L	0.08	7.7	311881	UCA	S	0.15	12.2	496448	UAA	*	0.30	1.0	40285	UGA	*	0.47	1.6	63237
UUG	L	0.13	12.9	525688	UCG	S	0.05	4.4	179419	UAG	*	0.24	0.8	32109	UGG	W	1.00	13.2	535595
CUU	L	0.13	13.2	536515	CCU	P	0.29	17.5	713233	CAU	H	0.42	10.9	441711	CGU	R	0.08	4.5	184609
CUC	L	0.20	19.6	796638	CCC	P	0.32	19.8	804620	CAC	H	0.58	15.1	613713	CGC	R	0.18	10.4	423516
CUA	L	0.07	7.2	290751	CCA	P	0.28	16.9	688038	CAA	Q	0.27	12.3	501911	CGA	R	0.11	6.2	250760
CUG	L	0.40	39.6	1611801	CCG	P	0.11	6.9	281570	CAG	Q	0.73	34.2	1391973	CGG	R	0.20	11.4	464485
AUU	I	0.36	16.0	650473	ACU	T	0.25	13.1	533609	AAU	N	0.47	17.0	689701	AGU	S	0.15	12.1	493429
AUC	I	0.47	20.8	846466	ACC	T	0.36	18.9	768147	AAC	N	0.53	19.1	776603	AGC	S	0.24	19.5	791383
AUA	I	0.17	7.5	304565	ACA	T	0.28	15.1	614523	AAA	K	0.43	24.4	993621	AGA	R	0.21	12.2	494682
AUG	M	1.00	22.0	896005	ACG	T	0.11	6.1	246105	AAG	K	0.57	31.9	1295568	AGG	R	0.21	12.0	486463
GUU	V	0.18	11.0	448607	GCU	A	0.27	18.4	750096	GAU	D	0.46	21.8	885429	GGU	G	0.16	10.8	437126
GUC	V	0.24	14.5	588138	GCC	A	0.40	27.7	1127679	GAC	D	0.54	25.1	1020595	GGC	G	0.34	22.2	903565
GUA	V	0.12	7.1	287712	GCA	A	0.23	15.8	643471	GAA	E	0.42	29.0	1177632	GGA	G	0.25	16.5	669873
GUG	V	0.46	28.1	1143534	GCG	A	0.11	7.4	299495	GAG	E	0.58	39.6	1609975	GGG	G	0.25	16.5	669768

Variation

While slight variations on the standard code had been predicted earlier, none were discovered until 1979, when researchers studying human mitochondrial genes discovered they used an alternative code. Many slight variants have been discovered since then, including various alternative mitochondrial codes, and small variants such as translation of the codon UGA as tryptophan in *Mycoplasma* species, and translation of CUG as a serine rather than a leucine in yeasts of the "CTG clade" (*Candida albicans* is member of this group). Because viruses must use the same genetic code as their hosts, modifications to the standard genetic code could interfere with the synthesis or functioning of viral proteins. However, some viruses (such as totiviruses) have adapted to the genetic code modification of the host. In bacteria and archaea, GUG and UUG are common start codons, but in rare cases, certain proteins may use alternative start codons not normally used by that species.

In certain proteins, non-standard amino acids are substituted for standard stop codons, depending on associated signal sequences in the messenger RNA. For example, UGA can code for selenocysteine and UAG can code for pyrrolysine. Selenocysteine is now viewed as the 21st amino acid, and pyrrolysine is viewed as the 22nd. Unlike selenocysteine, pyrrolysine encoded UAG is translated with the participation of a dedicated aminoacyl-tRNA synthetase. Both selenocysteine and pyrrolysine may be present in the same organism. Although the genetic code is normally fixed in an organism, the achaeal prokaryote *Acetohalobium arabaticum* can expand its genetic code from 20 to 21 amino acids (by including pyrrolysine) under different conditions of growth.

Despite these differences, all known naturally-occurring codes are very similar to each other, and the coding mechanism is the same for all organisms: three-base codons, tRNA, ribosomes, reading the code in the same direction and translating the code three letters at a time into sequences of amino acids.

Genetic code logo of the *Globobulimina pseudospinescens* mitochondrial genome. The logo shows the 64 codons from left to right, predicted alternatives in red (relative to the standard genetic code). Red line: stop codons. The height of each amino acid in the stack shows how often it is aligned to the codon in h omologous protein domains. The stack height indicates the support for the prediction.

Variant genetic codes used by an organism can be inferred by identifying highly conserved genes encoded in that genome, and comparing its codon usage to the amino acids in homologous proteins of other organisms. For example, the program FACIL infers a genetic code by searching which amino acids in homologous protein domains are most often aligned to every codon. The resulting amino acid probabilities for each codon are displayed in a genetic code logo, that also shows the support for a stop codon.

RNA Codon Table

nonpolar	polar	basic	acidic	(stop codon)

1st base	2nd base				3rd base
	Standard genetic code				
	U	C	A	G	
U	UUU (Phe/F) Phenylalanine	UCU (Ser/S) Serine	UAU (Tyr/Y) Tyrosine	UGU (Cys/C) Cysteine	U
	UUC	UCC	UAC	UGC	C
	UUA (Leu/L) Leucine	UCA	UAA Stop (Ochre)	UCA Stop (Opal)	A
	UUG	UCG	UAG Stop (Amber)	UGG (Trp/W) Tryptophan	G
C	CUU (Leu/L) Leucine	CCU (Pro/P) Proline	CAU (His/H) Histidine	CGU (Arg/R) Arginine	U
	CUC	CCC	CAC	CGC	C
	CUA	CCA	CAA (Gln/Q) Glutamine	CGA	A
	CUG	CCG	CAG	CGG	G
A	AUU (Ile/I) Isoleucine	ACU (Thr/T) Threonine	AAU (Asn/N) Asparagine	AGU (Ser/S) Serine	U
	AUC	ACC	AAC	AGC	C
	AUA	ACA	AAA (Lys/K) Lysine	AGA (Arg/R) Arginine	A
	AUG[A] (Met/M) Methionine	ACG	AAG	AGG	G
G	GUU (Val/V) Valine	GCU (Ala/A) Alanine	GAU (Asp/D) Aspartic acid	GGU (Gly/G) Glycine	U
	GUC	GCC	GAC	GGC	C
	GUA	GCA	GAA (Glu/E) Glutamic acid	GGA	A
	GUG	GCG	GAG	GGG	G

The codon AUG both codes for methionine and serves as an initiation site: the first AUG in an mRNA's coding region is where translation into protein begins.

Inverse table (compressed using IUPAC notation)					
Amino acid	Codons	Compressed	Amino acid	Codons	Compressed
Ala/A	GCU, GCC, GCA, GCG	GCN	Leu/L	UUA, UUG, CUU, CUC, CUA, CUG	YUR, CUN
Arg/R	CGU, CGC, CGA, CGG, AGA, AGG	CGN, MGR	Lys/K	AAA, AAG	AAR
Asn/N	AAU, AAC	AAY	Met/M	AUG	
Asp/D	GAU, GAC	GAY	Phe/F	UUU, UUC	UUY
Cys/C	UGU, UGC	UGY	Pro/P	CCU, CCC, CCA, CCG	CCN
Gln/Q	CAA, CAG	CAR	Ser/S	UCU, UCC, UCA, UCG, AGU, AGC	UCN, AGY
Glu/E	GAA, GAG	GAR	Thr/T	ACU, ACC, ACA, ACG	ACN
Gly/G	GGU, GGC, GGA, GGG	GGN	Trp/W	UGG	
His/H	CAU, CAC	CAY	Tyr/Y	UAU, UAC	UAY
Ile/I	AUU, AUC, AUA	AUH	Val/V	GUU, GUC, GUA, GUG	GUN
START	AUG		STOP	UAA, UGA, UAG	UAR, URA

DNA Codon Table

The DNA codon table is essentially identical to that for RNA, but with U replaced by T.

Origin

The origin of the genetic code is a part of the question of the origin of life. Under the main hypothesis for the origin of life, the RNA world hypothesis, any model for the emergence of genetic code is intimately related to a model of the transfer from ribozymes (RNA enzymes) to proteins as the principal enzymes in cells. In line with the RNA world hypothesis, transfer RNA molecules appear to have evolved before modern aminoacyl-tRNA synthetases, so the latter cannot be part of the explanation of its patterns.

A consideration of a hypothetical random genetic code further motivates a biochemical or evolutionary model for the origin of the genetic code. If amino acids were randomly assigned to triplet codons, there would be 1.5×10^{84} possible genetic codes to choose from. This number is found by calculating how many ways there are to place 21 items (20 amino acids plus one stop) in 64 bins, wherein each item is used at least once. In fact, the distribution of codon assignments in the genetic code is nonrandom. In particular, the genetic code clusters certain amino acid assignments. For example, amino acids that share the same biosynthetic pathway tend to have the same first base in their codons. This could be an evolutionary relic of early simpler genetic code with fewer amino acids, that later diverged to code for a larger set of amino acids. It could also reflect steric and chemical properties that had another effect on the codon during its evolution. Amino acids with similar physical properties also tend to have similar codons, reducing the problems caused by point mutations and mistranslations.

Given the non-random genetic triplet coding scheme, it has been suggested that a tenable hypothesis for the origin of genetic code should address multiple aspects of the codon table such as absence of codons for D-amino acids, secondary codon patterns for some amino acids, confinement of synonymous positions to third position, a limited set of only 20 amino acids instead of a number closer to 64, and the relation of stop codon patterns to amino acid coding patterns.

There are three main ideas for the origin of the genetic code, and many models belong to either one of them or to a combination thereof:

1. "Frozen accident": the genetic code has been randomly created. For example, early tRNA-like ribozymes may have had different affinities for amino acids, with codons emerging from another part of the ribozyme which exhibited random variability. Once enough peptides were coded for, any major random change in the genetic code would have been lethal, hence it is "frozen".

2. Stereochemical affinity: the genetic code is a result of a high affinity between each amino acid and its codon or anti-codon; the latter option implies that pre-tRNA molecules matched their corresponding amino acids by this affinity. Later during evolution, this matching has been gradually replaced with the one performed today by aminoacyl-tRNA synthetases.

3. Optimality: the genetic code continued to evolve after its initial creation, so that the current code reflects adaptation that maximizes some fitness function, usually some kind of error minimization.

Hypotheses for the origin of the genetic code have addressed a variety of scenarios:

- Chemical principles govern specific RNA interaction with amino acids. Experiments with aptamers showed that some amino acids have a selective chemical affinity for the base triplets that code for them. Recent experiments show that of the 8 amino acids tested, 6 show some RNA triplet-amino acid association.

- Biosynthetic expansion. The standard modern genetic code grew from a simpler earlier code through a process of "biosynthetic expansion". Here the idea is that primordial life "discovered" new amino acids (for example, as by-products of metabolism) and later incorporated some of these into the machinery of genetic coding. Although much circumstantial evidence has been found to suggest that fewer different amino acids were used in the past than today, precise and detailed hypotheses about which amino acids entered the code in what order have proved far more controversial.

- Natural selection has led to codon assignments of the genetic code that minimize the effects of mutations. A recent hypothesis suggests that the triplet code was derived from codes that used longer than triplet codons (such as quadruplet codons). Longer than triplet decoding would have higher degree of codon redundancy and would be more error resistant than the triplet decoding. This feature could allow accurate decoding in the absence of highly complex translational machinery such as the ribosome and before cells began making ribosomes.

- Information channels: Information-theoretic approaches model the process of translating the genetic code into corresponding amino acids as an error-prone information channel. The inherent noise (that is, the error) in the channel poses the organism with a fundamental question: how can a genetic code be constructed to withstand the effect of noise while accurately and efficiently translating information? These "rate-distortion" models suggest that the genetic code originated as a result of the interplay of the three conflicting evolutionary forces: the needs for diverse amino-acids, for error-tolerance and for minimal cost of resources. The code emerges at a coding transition when the mapping of codons to amino-acids becomes nonrandom. The emergence of the code is governed by the topology defined by the probable errors and is related to the map coloring problem.

- Game theory: Models based on signaling games combine elements of game theory, natural selection and information channels. Such models have been used to suggest that the first polypeptides were likely short and had some use other than enzymatic function. Game theoretic models have also suggested that the organization of RNA strings into cells may have been necessary to prevent "deceptive" use of the genetic code, i.e. preventing the ancient equivalent of viruses from overwhelming the RNA world.

- Stop codons: Codons for translational stops are also an interesting aspect to the problem of the origin of the genetic code. As an example for addressing stop codon evolution, it has been suggested that the stop codons are such that they are most likely to terminate translation early in the case of a frame shift error. In contrast, some stereochemical molecular models explain the origin of stop codons as "unassignable".

Expanded Genetic Codes (Synthetic Biology)

Since 2001, 40 non-natural amino acids have been added into protein by creating a unique codon (recoding) and a corresponding transfer-RNA:aminoacyl − tRNA-synthetase pair to encode it with diverse physicochemical and biological properties in order to be used as a tool to exploring protein structure and function or to create novel or enhanced proteins.

H. Murakami and M. Sisido have extended some codons to have four and five bases. Steven A. Benner constructed a functional 65th (*in vivo*) codon.

Transfer RNA

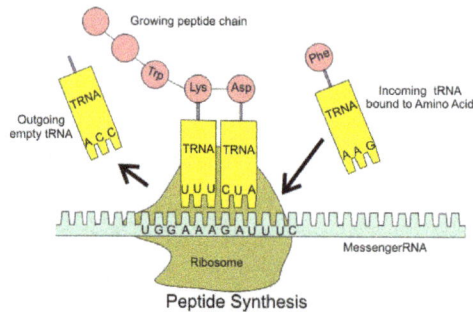

The interaction of tRNA and mRNA in protein synthesis.

A transfer RNA (abbreviated tRNA and formerly referred to as sRNA, for soluble RNA) is an adaptor molecule composed of RNA, typically 76 to 90 nucleotides in length, that serves as the physical link between the mRNA and the amino acid sequence of proteins. It does this by carrying an amino acid to the protein synthetic machinery of a cell (ribosome) as directed by a three-nucleotide sequence (codon) in a messenger RNA (mRNA). As such, tRNAs are a necessary component of translation, the biological synthesis of new proteins in accordance with the genetic code.

Overview

While the specific nucleotide sequence of an mRNA specifies which amino acids are incorporated into the protein product of the gene from which the mRNA is transcribed, the role of *tRNA* is to specify which sequence from the genetic code corresponds to which amino acid. The mRNA encodes a protein as a series of contiguous codons, each of which is recognized by a particular tRNA. One end of the tRNA matches the genetic code in a three-nucleotide sequence called the anticodon. The anticodon forms three base pairs with a codon in mRNA during protein biosynthesis. On the other end of the tRNA is a covalent attachment to the amino acid that corresponds to the anticodon sequence. Each type of tRNA molecule can be attached to only one type of amino acid, so each organism has many types of tRNA. Because the genetic code contains multiple codons that specify the same amino acid, there are several tRNA molecules bearing different anticodons which carry the same amino acid.

The covalent attachment to the tRNA 3' end is catalyzed by enzymes called aminoacyl tRNA synthetases. During protein synthesis, tRNAs with attached amino acids are delivered to the ribosome by proteins called elongation factors, which aid in association of the tRNA with the ribosome, synthesis of the new polypeptide and translocation (movement) of the ribosome along the mRNA. If the tRNA's anticodon matches the mRNA, another tRNA already bound to the ribosome transfers the growing polypeptide chain from its 3' end to the amino acid attached to the 3' end of the newly delivered tRNA, a reaction catalyzed by the ribosome. A large number of the individual nucleotides in a tRNA molecule may be chemically modified, often by methylation or deam-idation. These unusual bases sometimes affect the tRNA's interaction with ribosomes and sometimes occur in the anticodon to alter base-pairing properties.

Structure

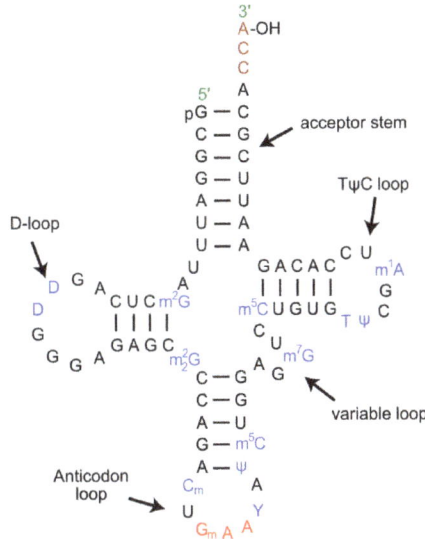

Secondary *cloverleaf structure* of tRNA^{Phe} from yeast.

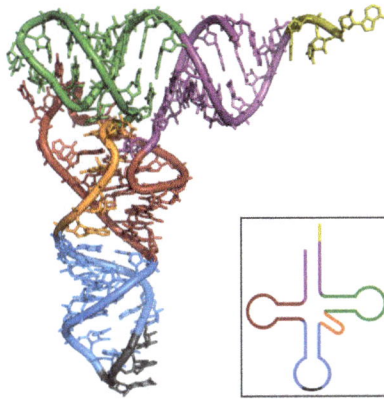

Tertiary structure of tRNA. *CCA tail* in yellow, *Acceptor stem* in purple, *Variable loop* in orange, *D arm* in red, *Anticodon arm* in blue with *Anticodon* in black, *T arm* in green.

The structure of tRNA can be decomposed into its primary structure, its secondary structure (usually visualized as the *cloverleaf structure*), and its tertiary structure (all tRNAs have a similar L-shaped 3D structure that allows them to fit into the P and A sites of the ribosome). The cloverleaf structure becomes the 3D L-shaped structure through coaxial stacking of the helices, which is a common RNA tertiary structure motif.

The lengths of each arm, as well as the loop 'diameter', in a tRNA molecule vary from species to species.

The tRNA structure consists of the following:

1. A 5'-terminal phosphate group.

2. The acceptor stem is a 7- to 9-base pair (bp) stem made by the base pairing of the 5'-terminal nucleotide with the 3'-terminal nucleotide (which contains the CCA 3'-terminal group used to attach the amino acid). The acceptor stem may contain non-Watson-Crick base pairs.

3. The CCA tail is a cytosine-cytosine-adenine sequence at the 3' end of the tRNA molecule. The amino acid loaded onto the tRNA by aminoacyl tRNA synthetases, to form aminoacyl-tRNA, is covalently bonded to the 3'-hydroxyl group on the CCA tail. This sequence is important for the recognition of tRNA by enzymes and critical in translation. In prokaryotes, the CCA sequence is transcribed in some tRNA sequences. In most prokaryotic tRNAs and eukaryotic tRNAs, the CCA sequence is added during processing and therefore does not appear in the tRNA gene.

4. The D arm is a 4- to 6-bp stem ending in a loop that often contains dihydrouridine.

5. The anticodon arm is a 5-bp stem whose loop contains the anticodon. The tRNA 5'-to-3' primary structure contains the anticodon but in reverse order, since 3'-to-5' directionality is required to read the mRNA from 5'-to-3'.

6. The T arm is a 4- to 5- bp stem containing the sequence TΨC where Ψ is pseudouridine, a modified uridine.

7. Bases that have been modified, especially by methylation (e.g. tRNA (guanine-N7-)-methyltransferase), occur in several positions throughout the tRNA. The first anticodon base, or wobble-position, is sometimes modified to inosine (derived from adenine), pseudouridine or lysidine (derived from cytosine).

Anticodon

An anticodon is a unit made up of three nucleotides that correspond to the three bases of the codon on the mRNA. Each tRNA contains a distinct anticodon triplet sequence that can base-pair to one or more codons for an amino acid. Some anticodons can pair with more than one codon due to a phenomenon known as wobble base pairing. Frequently, the first nucleotide of the anticodon is one not found on mRNA: inosine, which can hydrogen bond to more than one base in the corresponding codon position. In the genetic code, it is common for a single amino acid to be specified by all four third-position possibilities, or at least by both pyrimidines and purines; for example, the amino acid glycine is coded for by the codon sequences GGU, GGC, GGA, and GGG. Other modified nucleotides may also appear at the first anticodon position - sometimes known as the "wobble position" - resulting in subtle changes to the genetic code, as for example in mitochondria.

To provide a one-to-one correspondence between tRNA molecules and codons that specify amino acids, 61 types of tRNA molecules would be required per cell. However, many cells contain fewer than 61 types of tRNAs because the wobble base is capable of binding to several, though not necessarily all, of the codons that specify a particular amino acid. A minimum of 31 tRNA are required to translate, unambiguously, all 61 sense codons of the standard genetic code.

Aminoacylation

Aminoacylation is the process of adding an aminoacyl group to a compound. It covalently links an amino acid to the CCA 3' end of a tRNA molecule.

Each tRNA is aminoacylated (or *charged*) with a specific amino acid by an aminoacyl tRNA synthetase. There is normally a single aminoacyl tRNA synthetase for each amino acid, despite the fact that there can be more than one tRNA, and more than one anticodon, for an amino acid. Recognition of the appropriate tRNA by the synthetases is not mediated solely by the anticodon, and the acceptor stem often plays a prominent role.

Reaction:

1. amino acid + ATP → aminoacyl-AMP + PPi

2. aminoacyl-AMP + tRNA → aminoacyl-tRNA + AMP

Certain organisms can have one or more aminoacyl tRNA synthetases missing. This leads to charging of the tRNA by a chemically related amino acid. An enzyme or enzymes modify the charged amino acid to the final one. For example, *Helicobacter pylori* has glutaminyl tRNA synthetase missing. Thus, glutamate tRNA synthetase charges tRNA-glutamine(tRNA-Gln) with glutamate. An amidotransferase then converts the acid side chain of the glutamate to the amide, forming the correctly charged gln-tRNA-Gln.

Binding to Ribosome

The range of conformations adopted by tRNA as it transits the A/T through P/E sites on the ribosome. The Protein Data Bank (PDB) codes for the structural models used as end points of the animation are given. Both tRNAs are modeled as phenylalanine-specific tRNA from *Escherichia coli*, with the A/T tRNA as a homology model of the deposited coordinates. Color coding as shown for tRNA tertiary structure. Adapted from.

The ribosome has three binding sites for tRNA molecules that span the space between the two ribosomal subunits: the A (aminoacyl), P (peptidyl), and E (exit) sites. In addition, the ribosome has two other sites for tRNA binding that are used during mRNA decoding or during the initiation of protein synthesis. These are the T site (named elongation factor Tu) and I site (initiation). By convention, the tRNA binding sites are denoted with the site on the small ribosomal subunit listed first and the site on the large ribosomal subunit listed second. For example, the A site is often written A/A, the P site,

P/P, and the E site, E/E. The binding proteins like L27, L2, L14, L15, L16 at the A- and P- sites have been determined by affinity labeling by A.P. Czernilofsky et al. (*Proc. Natl. Acad. Sci, USA*, pp 230–234, 1974).

Once translation initiation is complete, the first aminoacyl tRNA is located in the P/P site, ready for the elongation cycle described below. During translation elongation, tRNA first binds to the ribosome as part of a complex with elongation factor Tu (EF-Tu) or its eukaryotic (eEF-1) or archaeal counterpart. This initial tRNA binding site is called the A/T site. In the A/T site, the A-site half resides in the small ribosomal subunit where the mRNA decoding site is located. The mRNA decoding site is where the mRNA codon is read out during translation. The T-site half resides mainly on the large ribosomal subunit where EF-Tu or eEF-1 interacts with the ribosome. Once mRNA decoding is complete, the aminoacyl-tRNA is bound in the A/A site and is ready for the next peptide bond to be formed to its attached amino acid. The peptidyl-tRNA, which transfers the growing polypeptide to the aminoacyl-tRNA bound in the A/A site, is bound in the P/P site. Once the peptide bond is formed, the tRNA in the P/P site is deacylated, or has a free 3' end, and the tRNA in the A/A site carries the growing polypeptide chain. To allow for the next elongation cycle, the tRNAs then move through hybrid A/P and P/E binding sites, before completing the cycle and residing in the P/P and E/E sites. Once the A/A and P/P tRNAs have moved to the P/P and E/E sites, the mRNA has also moved over by one codon and the A/T site is vacant, ready for the next round of mRNA decoding. The tRNA bound in the E/E site then leaves the ribosome.

The P/I site is actually the first to bind to aminoacyl tRNA, which is delivered by an initiation factor called IF2 in bacteria. However, the existence of the P/I site in eukaryotic or archaeal ribosomes has not yet been confirmed. The P-site protein L27 has been determined by affinity labeling by E. Collatz and A.P. Czernilofsky (*FEBS Lett.*, Vol. 63, pp 283–286, 1976).

tRNA Genes

Organisms vary in the number of tRNA genes in their genome. The nematode worm *C. elegans*, a commonly used model organism in genetics studies, has 29,647 genes in its nuclear genome, of which 620 code for tRNA. The budding yeast *Saccharomyces cerevisiae* has 275 tRNA genes in its genome.

In the human genome, which, according to January 2013 estimates, has about 20,848 protein coding genes in total, there are 497 nuclear genes encoding cytoplasmic tRNA molecules, and 324 tRNA-derived pseudogenes—tRNA genes thought to be no longer functional (although pseudo tRNAs have been shown to be involved in antibiotic resistance in bacteria). Regions in nuclear chromosomes, very similar in sequence to mitochondrial tRNA genes, have also been identified (tRNA-lookalikes). These tRNA-lookalikes are also considered part of the nuclear mitochondrial DNA (genes transferred from the mitochondria to the nucleus).

As with all eukaryotes, there are 22 mitochondrial tRNA genes in humans. Mutations in some of these genes have been associated with severe diseases like the MELAS syndrome.

Cytoplasmic tRNA genes can be grouped into 49 families according to their anticodon features. These genes are found on all chromosomes, except 22 and Y chromosome. High clustering on 6p is observed (140 tRNA genes), as well on 1 chromosome.

The HGNC, in collaboration with the Genomic tRNA Database (GtRNAdb) and experts in the field, has approved unique names for human genes that encode tRNAs.

Evolution

Genomic tRNA content is a differentiating feature of genomes among biological domains of life: Archaea present the simplest situation in terms of genomic tRNA content with a uniform number of gene copies, Bacteria have an intermediate situation and Eukarya present the most complex situation. Eukarya present not only more tRNA gene content than the other two kingdoms but also a high variation in gene copy number among different isoacceptors, and this complexity seem to be due to duplications of tRNA genes and changes in anticodon specificity.

Evolution of the tRNA gene copy number across different species has been linked to the appearance of specific tRNA modification enzymes (uridine methyltransferases in Bacteria, and adenosine deaminases in Eukarya), which increase the decoding capacity of a given tRNA. As an example, tRNAAla encodes four different tRNA isoacceptors (AGC, UGC, GGC and CGC). In Eukarya, AGC isoacceptors are extremely enriched in gene copy number in comparison to the rest of isoacceptors, and this has been correlated with its A-to-I modification of its wobble base. This same trend has been shown for most amino acids of eukaryal species. Indeed, the effect of these two tRNA modifications is also seen in codon usage bias. Highly expressed genes seem to be enriched in codons that are exclusively using codons that will be decoded by these modified tRNAs, which suggests a possible role of these codons—and consequently of these tRNA modifications—in translation efficiency.

tRNA-derived Fragments

tRNA-derived fragments (or tRFs) are short molecules that emerge after cleavage of the mature tRNAs or the precursor transcript. Both cytoplasmic and mitochondrial tRNAs can produce fragments. There are at least four structural types of tRFs believed to originate from mature tRNAs, including the relatively long tRNA halves and short 5'-tRFs, 3'-tRFs and i-tRFs. The precursor tRNA can be cleaved to produce molecules from the 5' leader or 3' trail sequences. Cleavage enzymes include Angiogenin, Dicer, RNase Z and RNase P. Especially in the case of Angiogenin, the tRFs have a characteristically unusual cyclic phosphate at their 3' end and a hydroxyl group at the 5' end.

tRFs have multiple dependencies and roles. They exhibit significant changes between sexes, among races and disease status. Functionally, they can be loaded on Ago and act through RNAi pathways, participate in the formation of stress granules, displace mRNAs from RNA-binding proteins or inhibit translation. At the system or the organismal level, the four types of tRFs have a diverse spectrum of activities. Functionally, tRFs are associated with viral infection, cancer, cell proliferation and also with epigenetic transgenerational regulation of metabolism.

tRFs are not restricted to humans but have been shown to exist in multiple organisms.

Two online tools are available for those wishing to learn more about tRFs: the framework for the interactive exploration of mitochondrial and nuclear tRNA fragments (MINTbase) and the relational database of Transfer RNA related Fragments(tRFdb),. MINTbase also provides a naming scheme for the naming of tRFs called tRF-license plates that is genome independent.

tRNA Biogenesis

In eukaryotic cells, tRNAs are transcribed by RNA polymerase III as pre-tRNAs in the nucleus. RNA polymerase III recognizes two highly conserved downstream promoter sequences: the 5' intragenic control region (5'-ICR, D-control region, or A box), and the 3'-ICR (T-control region or B box) inside tRNA genes. The first promoter begins at +8 of mature tRNAs and the second promoter is located 30-60 nucleotides downstream of the first promoter. The transcription terminates after a stretch of four or more thymidines.

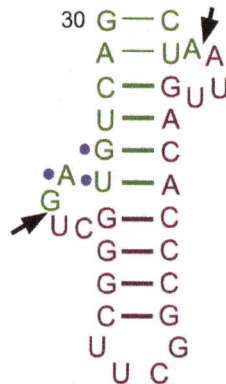

Bulge-helix-bulge motive of tRNA intron

Pre-tRNAs undergo extensive modifications inside the nucleus. Some pre-tRNAs contain introns that are spliced, or cut, to form the functional tRNA molecule; in bacteria these self-splice, whereas in eukaryotes and archaea they are removed by tRNA-splicing endonucleases. Eukaryotic pre-tRNA contains bulge-helix-bulge (BHB) structure motif that is important for recognition and precise splicing of tRNA intron by endonucleases. This motif position and structure are evolutionary conserved. However, some organisms, such as unicellular algae have a non-canonical position of BHB-motif as well as 5'- and 3'-ends of the spliced intron sequence. The 5' sequence is removed by RNase P, whereas the 3' end is removed by the tRNase Z enzyme. A notable exception is in the archaeon *Nanoarchaeum equitans,* which does not possess an RNase P enzyme and has a promoter placed such that transcription starts at the 5' end of the mature tRNA. The non-templated 3' CCA tail is added by a nucleotidyl transferase. Before tRNAs are exported into the cytoplasm by Los1/Xpo-t, tRNAs are aminoacylated. The order of the processing events is not conserved. For example, in yeast, the splicing is not carried out in the nucleus but at the cytoplasmic side of mitochondrial membranes.

History

The existence of tRNA was first hypothesized by Francis Crick, based on the assumption that there must exist an adapter molecule capable of mediating the translation of the RNA alphabet into the protein alphabet. Significant research on structure was conducted in the early 1960s by Alex Rich and Don Caspar, two researchers in Boston, the Jacques Fresco group in Princeton University and a United Kingdom group at King's College London. In 1965, Robert W. Holley of Cornell University reported the primary structure and suggested three secondary structures. tRNA was first crystallized in Madison, Wisconsin, by Robert M. Bock. The cloverleaf structure was ascertained by several other studies in the following years and was finally confirmed using X-ray crystallography studies in 1974. Two independent groups, Kim Sung-Hou working under Alexander Rich and a British group headed by Aaron Klug, published the same crystallography findings within a year.

Stop Codon

In the genetic code, a stop codon (or termination codon) is a nucleotide triplet within messenger RNA that signals a termination of translation into proteins. Proteins are based on polypeptides, which are unique sequences of amino acids. Most codons in messenger RNA (from DNA) correspond to the addition of an amino acid to a growing polypeptide chain, which may ultimately become a protein. Stop codons signal the termination of this process by binding release factors, which cause the ribosomal subunits to disassociate, releasing the amino acid chain. While start codons need nearby sequences or initiation factors to start translation, a stop codon alone is sufficient to initiate termination.

Introduction

In the standard genetic code, there are three different stop codons:

- in RNA:

- UAG ("amber")

- UAA ("ochre")

- UGA ("opal")

- in DNA:

  - TAG ("amber")

  - TAA ("ochre")

  - TGA ("opal" or "umber")

In 2007, the UGA codon has been identified as the codon coding for Selenocysteine (Sec). This amino acid is found in 25 selenoproteins where it is located in the active site of the protein. Transcription of this codon is enabled by the proximity of the SECIS element (SElenoCysteine Incorporation Sequence). The UAG codon can translate into pyrrolysine in a similar manner.

Distribution of stop codons within the genome of an organism is non-random and can correlate with GC-content. For example, the *E. coli* K-12 genome contains 2705 TAA (63%), 1257 TGA (29%), and 326 TAG (8%) stop codons (GC content 50.8%). Also the substrates for the stop codons release factor 1 or release factor 2 are strongly correlated to the abundance of stop codons. Large scale study of bacteria with a broad range of GC-contents shows that while the frequency of occurrence of TAA is negatively correlated to the GC-content and the frequency of occurrence of TGA is positively correlated to the GC-content, the frequency of occurrence of the TAG stop codon, which is often the minimally used stop codon in a genome, is not influenced by the GC-content.

Nonsense mutations are changes in DNA sequence that introduce a premature stop codon, causing any resulting protein to be abnormally shortened. This often causes a loss of function in the protein, as critical parts of the amino acid chain are no longer created. Because of this terminology, stop codons have also been referred to as nonsense codons.

Amber, Ochre, and Opal Nomenclature

Stop codons were historically given many different names, as they each corresponded to a distinct class of mutants that all behaved in a similar manner. These mutants were first isolated within bacteriophages (T4 and lambda), viruses that infect the bacteria *Escherichia coli*. Mutations in viral genes weakened their infectious ability, sometimes creating viruses that were able to infect and grow within only certain varieties of *E coli*.

Amber Mutations (Uag)

> were the first set of nonsense mutations to be discovered, isolated by Richard Epstein and Charles Steinberg and named after their friend Harris Bernstein (whose last name means "amber" in German).

Viruses with amber mutations are characterized by their ability to infect only certain strains of bacteria, known as amber suppressors. These bacteria carry their own mutation that allows a recovery of function in the mutant viruses. For example, a mutation in the tRNA that recognizes the amber stop codon allows translation to "read through" the codon and produce a full-length protein, thereby recovering the normal form of the protein and "suppressing" the amber mutation.

Thus, amber mutants are an entire class of virus mutants that can grow in bacteria that contain amber suppressor mutations. Similar suppressors are known for ochre and opal stop codons as well.

Ochre Mutation (Uaa)

was the second stop codon mutation to be discovered. Given a color name to match the name of amber mutants, ochre mutant viruses had a similar property in that they recovered infectious ability within certain suppressor strains of bacteria. The set of ochre suppressors was distinct from amber suppressors, so ochre mutants were inferred to correspond to a different nucleotide triplet. Through a series of mutation experiments comparing these mutants with each other and other known amino acid codons, Sydney Brenner concluded that the amber and ochre mutations corresponded to the nucleotide triplets "UAG" and "UAA".

opal mutations or umber mutations (UGA)

the third and last stop codon in the standard genetic code was discovered soon after, corresponding to the nucleotide triplet "UGA". Nonsense mutations that created this premature stop codon were later called *opal mutations* or *umber mutations*.

Hidden Stops

Hidden stops are non-stop codons that would be read as stop codons if they were frameshifted +1 or -1. These prematurely terminate translation if the corresponding frame-shift (such as due to a ribosomal RNA slip) occurs before the hidden stop. It is hypothesised that this decreases resource waste on nonfunctional proteins and the production of potential cytotoxins. Researchers at Louisiana State University propose the *ambush hypothesis*, that hidden stops are selected for. Codons that can form hidden stops are used in genomes more frequently compared to synonymous codons that would otherwise code for the same amino acid. Unstable rRNA in an organism correlates with a higher frequency of hidden stops. This hypothesis however could not be validated with a larger data set.

Stop-codons and hidden stops together are collectively referred as stop-signals. Researchers at University of Memphis found that the ratios of the stop-signals on the three reading frames of a genome (referred to as translation stop-signals ratio or TSSR) of genetically related bacteria, despite their great differences in gene contents, are much alike. This nearly identical Genomic-TSSR value of genetically related bacteria may suggest that bacterial genome expansion is limited by their unique stop-signals bias of that bacterial species.

Translational Readthrough

Stop codon suppression or translational readthrough occurs when in translation a stop codon is interpreted as a sense codon, that is, when a (standard) amino acid is 'encoded' by the stop codon. Mutated tRNAs can be the cause of readthrough, but also certain nucleotide motifs close to the stop codon. Translational readthrough is very common in viruses and bacteria, and has also been found as a gene regulatory principle in humans.

Nonstop Mutations

A nonstop mutation is a point mutation that occurs within a stop codon. Nonstop mutations cause the continued translation of an mRNA strand into an untranslated region. Most polypeptides resulting from a gene with a nonstop mutation are nonfunctional due to their extreme length. Nonstop mutations differ from nonsense mutations in that they do not create a stop codon but, instead, delete one.

Nonstop mutations have been linked with several congenital diseases including congenital adrenal hyperplasia, variable anterior segment dysgenesis, and mitochondrial neurogastrointestinal encephalomyopathy.

Use as a Watermark

When Craig Venter unveiled the first fully functioning, reproducing cell controlled by synthetic DNA he described how his team used frequent stop codons to create watermarks in RNA and DNA to help confirm the results were indeed synthetic (and not contaminated or otherwise), using it to encode authors names and website addresses.

References

- Alberts, et. al, Bruce (2008). Molecular Biology of the Cell (5th ed.). New York: Garland Science. ISBN 0-8153-4105-9.

- Taylor, Stanley R. Maloy, Valley J. Stewart, Ronald K. (1996). Genetic analysis of pathogenic bacteria : a laboratory manual. New York: Cold Spring Harbor Laboratory. ISBN 978-0-87969-453-1.

- Crick F (1988). "Chapter 8: The genetic code". What mad pursuit: a personal view of scientific discovery. New York: Basic Books. pp. 89–101. ISBN 0-465-09138-5.

- Mulligan PK, King RC, Stansfield WD (2006). A dictionary of genetics. Oxford [Oxfordshire]: Oxford University Press. p. 608. ISBN 0-19-530761-5.

- Griffiths AJ, Miller JH, Suzuki DT, Lewontin RC, et al., eds. (2000). "Spontaneous mutations". An Introduction to Genetic Analysis (7th ed.). New York: W. H. Freeman. ISBN 0-7167-3520-2.

- Lewis R (2005). Human Genetics: Concepts and Applications (6th ed.). Boston, Mass: McGraw Hill. pp. 227–228. ISBN 0-07-111156-5.

- Watson JD, Baker TA, Bell SP, Gann A, Levine M, Oosick R (2008). Molecular Biology of the Gene. San Francisco: Pearson/Benjamin Cummings. ISBN 0-8053-9592-X.

- Yang et al. (1990) in Michel-Beyerle, M. E., ed. Reaction centers of photosynthetic bacteria: Feldafing-II-Meeting 6. Berlin: Springer-Verlag. pp. 209–18. ISBN 3-540-53420-2

- Yarus M (2010). Life from an RNA World: The Ancestor Within. Cambridge: Harvard University Press. p. 163. ISBN 0-674-05075-4.

- Simon M (2005). Emergent computation: emphasizing bioinformatics. New York: AIP Press/Springer Science+Business Media. pp. 105–106. ISBN 0-387-22046-1.

- Stryer L, Berg JM, Tymoczko JL (2002). Biochemistry (5th ed.). San Francisco: W.H. Freeman. ISBN 0-7167-4955-6.

Permissions

All chapters in this book are published with permission under the Creative Commons Attribution Share Alike License or equivalent. Every chapter published in this book has been scrutinized by our experts. Their significance has been extensively debated. The topics covered herein carry significant information for a comprehensive understanding. They may even be implemented as practical applications or may be referred to as a beginning point for further studies.

We would like to thank the editorial team for lending their expertise to make the book truly unique. They have played a crucial role in the development of this book. Without their invaluable contributions this book wouldn't have been possible. They have made vital efforts to compile up to date information on the varied aspects of this subject to make this book a valuable addition to the collection of many professionals and students.

This book was conceptualized with the vision of imparting up-to-date and integrated information in this field. To ensure the same, a matchless editorial board was set up. Every individual on the board went through rigorous rounds of assessment to prove their worth. After which they invested a large part of their time researching and compiling the most relevant data for our readers.

The editorial board has been involved in producing this book since its inception. They have spent rigorous hours researching and exploring the diverse topics which have resulted in the successful publishing of this book. They have passed on their knowledge of decades through this book. To expedite this challenging task, the publisher supported the team at every step. A small team of assistant editors was also appointed to further simplify the editing procedure and attain best results for the readers.

Apart from the editorial board, the designing team has also invested a significant amount of their time in understanding the subject and creating the most relevant covers. They scrutinized every image to scout for the most suitable representation of the subject and create an appropriate cover for the book.

The publishing team has been an ardent support to the editorial, designing and production team. Their endless efforts to recruit the best for this project, has resulted in the accomplishment of this book. They are a veteran in the field of academics and their pool of knowledge is as vast as their experience in printing. Their expertise and guidance has proved useful at every step. Their uncompromising quality standards have made this book an exceptional effort. Their encouragement from time to time has been an inspiration for everyone.

The publisher and the editorial board hope that this book will prove to be a valuable piece of knowledge for students, practitioners and scholars across the globe.

Index

A

Alkylation, 22-25, 68, 179

Alloprotein, 202

Alpha-aminobutyric Acid, 117

Amber Codon Suppression, 197

Amino Acid Dating, 35-36

Amino Acid Synthesis, 13, 15, 133, 135, 148, 168, 179-180

Aminoacylation, 217

Aminocaproic Acid, 115-117

Anticodon, 193, 214-217, 219

Arndt-eistert Reaction, 133, 158

Aromaticity, 47

Asymmetric Strecker Reactions, 182

Atherosclerosis, 70, 124

B

Bargellini Reaction, 159

Beta-alanine, 114

Biocatalysis, 185

Biochemical Pathways, 66-67

Biochemistry, 2, 4, 6-8, 10, 12, 14, 16, 18, 20, 22, 24, 26, 28, 30, 32, 34, 36, 38-40, 42, 44, 46-48, 50, 52, 54, 56, 58, 60, 62, 64, 66, 68, 70, 72, 74, 76, 78, 80, 82, 84-85, 88, 90, 92, 94, 96, 98, 100, 102, 104, 106, 108, 110, 112, 114, 116-118, 120, 122, 124, 126, 128, 130-134, 136, 138, 140, 142, 144, 146, 148, 150, 152, 154, 156, 158, 160, 162, 164, 166, 168, 170, 172, 174, 176, 178, 180, 182, 184, 186, 188, 190-191, 194, 196, 198, 200, 202, 204, 206, 208, 210, 212, 214, 216, 218, 220, 222, 224-225

Biodegradable Plastics, 3, 12

Biogenesis, 220

Biomolecules, 9, 56, 134, 142, 161

Biosynthesis, 2, 10, 14, 45, 47-50, 52, 56-58, 62, 64-66, 70-74, 76-79, 83-87, 90, 104, 107, 113-114, 117, 121, 123-124, 127, 129, 133-139, 141, 143, 188, 191, 214

Bone Mass, 124

Brain Development, 88, 119

C

Carbodiimides, 152-154

Carboxyl Radical, 30

Carboxylic Acid, 1-2, 5, 13, 15, 23, 27-28, 30-34, 45, 50, 52, 56, 62, 74, 77, 83, 86, 105-106, 113, 147, 151, 153, 158-159, 171

Carnitine, 2, 8, 70, 123-126

Catabolism, 8, 15, 51, 67, 103, 112, 121-123

Chemical Synthesis, 13, 68, 123, 133, 135, 137, 139, 141, 143, 145, 147, 149, 151, 153, 155, 157, 159, 161, 163, 165, 167, 169, 171, 173, 175, 177, 179, 181, 183, 185, 187-189, 191, 194, 203

Chiral Auxiliaries, 184-186, 190-191

Chiral Pool Synthesis, 12, 185-186

Citric Acid Cycle, 9, 16, 87, 121, 123, 133-135

Citrulline, 8, 44, 106, 112, 127-128

Codon Assignment, 195

Corey-link Reaction, 133, 159

Cyclic Peptides, 157

D

Degeneracy, 207

Diazotization, 24

Domoic Acid, 128-131

E

Enantiomers, 3, 163, 182-183, 186, 188-189

Enantioselective Organocatalysis, 185, 190

Enantioselective Synthesis, 174, 182-189

Eosinophilia, 83

Ethylene Synthesis, 67

G

Gamma-aminobutyric Acid, 8, 86, 88, 117, 121

Gene Expression, 90, 98, 138

Genetic Code, 1-2, 7-9, 12, 16, 63, 90-91, 104-105, 110-112, 164, 192-195, 197-205, 207, 209-217, 219, 221, 223-225

Glutamatergic Signaling Circuits, 88

Glutamic Acid, 2, 11, 14, 18, 36, 41, 57, 83, 86-87, 89, 93, 96-103, 110, 112, 155, 165, 207, 210

H

Histidine, 41-42, 44-50, 56, 90, 93, 96, 98-101, 106, 114-115, 133-134,

136, 139-140, 157, 210

I

In Vitro Synthesis, 203

Industrial Production, 77

Insulin Resistance, 55, 84

Interchangeability, 44

Isoelectric Point, 6-7

Isoleucine, 5, 18, 36, 41-42, 44, 50-52, 63, 83, 90, 93, 96, 98-102, 109-110, 117, 137, 139, 142, 166, 198, 203

Isomerism, 3

L

Leucine, 3, 5, 18, 36, 41-42, 44, 50-56, 83-84, 90, 94, 96, 98-102, 109-110, 112, 141-142, 166, 203, 206-207, 209

Liquid-phase Synthesis, 143, 148

Lysine, 4, 8, 11, 16, 26, 41-44, 56-62, 64-65, 90, 94, 96-102, 104-105, 107-109, 112, 116, 123, 137-139, 155, 198, 200, 205, 210

M

Mass Spectrometry, 37, 99, 164

Methionine, 11, 15, 41-44, 62-70, 90, 94, 96-100, 102, 105, 108, 110-111, 123, 135, 137-140, 166, 198, 202-203, 205-206, 210

Methionine Catabolism, 67

Myalgia Syndrome, 83

N

Nervous System, 74, 88, 117, 120-121

Neurotransmitter, 2, 4, 8, 10, 74, 78, 81, 86, 88, 106, 115, 117-119, 121, 130

Nitrogen Fixation, 134

Nomenclature, 5, 28-29, 39, 84, 92, 106, 222

Non-protein Functions, 10

Non-proteinogenic

Amino Acids, 2, 8, 91, 104, 111-113, 194

Nonstop Mutations, 224

Nullomers, 12, 37-39

Nutritional Sources, 51

O

On Resin Cyclization, 157

Opal Nomenclature, 222

Oxaloacetate, 15, 51, 87, 134, 137

P

Peptide Bond Formation, 13

Peptide Synthesis, 14, 133, 143-148, 150, 152, 154-157, 191-192

Petasis Reaction, 166-171, 173-174, 177-178

Pharmacology, 89, 122

Phenylalanine, 10-11, 36, 41-42, 44, 70-74, 90, 94, 96, 98-103, 134, 136-137, 158, 165, 177, 201, 204-205, 217

Phosphoenolpyruvate, 136

Phosphoglycerates, 140

Polyamide Resin, 147

Prebiotic Amino Acids, 109

Protein Sources, 42

Proteinogenic Amino Acid, 5, 62, 86, 90, 112, 180

Proteinogenic Amino Acids, 2, 4, 7-9, 15, 62, 75, 90-91, 104-107, 111-113, 123, 192, 194

R

Racemization, 35-37, 153, 155, 157, 173, 185

Recoded Synthetic Genome, 201

Reverse-transulfurylation Pathway, 67

Ribosome, 4, 7, 13, 63, 91, 140, 144, 193-194, 203-206, 213-215, 217-218

S

Side Chain Properties, 97

Side Chains, 4, 96, 101, 112, 152, 154, 157, 195

Solid-phase Synthesis, 143, 154, 157, 190

Stoichiometry, 100

Stop Codon, 2, 8-9, 90, 98-99, 105, 192-193, 195, 197, 206, 209-210, 212-213, 221-224

Strecker Amino Acid Synthesis, 13, 133, 180

Sulfonation, 24

Synthetic Biology, 192-193, 213

T

Tautomerism, 46

Threonine, 3, 11, 41-44, 47, 64-65, 74-76, 90, 95-100, 102, 110, 137-139

Thyroid Hormone, 125

Total Synthesis, 154, 178, 186, 188

Toxicology, 130

Trans-sulfurylation Pathway, 66

Transamination, 51, 59, 87-88, 133-135, 141-142

Translational Readthrough, 224

Tryptophan, 10-11, 41-44, 63, 71, 77-83, 90, 95, 97-103, 136-137, 139, 209-210

Twin Amino Acid Stereocentres, 108

Z

Zwitterions, 5-6